ZIML Math Competition Book

Division M 2017-2018

Areteem Institute

ZIML Math Competition Book Division M 2017-18

Edited by   John Lensmire
David Reynoso
Kevin Wang
Kelly Ren

ISBN: 1-944863-28-1
ISBN-13: 978-1-944863-28-9

First printing, November 2018.

TITLES PUBLISHED BY ARETEEM PRESS

Cracking the High School Math Competitions (and Solutions Manual) - Covering AMC 10 & 12, ARML, and ZIML

Mathematical Wisdom in Everyday Life (and Solutions Manual) - From Common Core to Math Competitions

Geometry Problem Solving for Middle School (and Solutions Manual) - From Common Core to Math Competitions

Fun Math Problem Solving For Elementary School (and Solutions Manual)

ZIML MATH COMPETITION BOOK SERIES

ZIML Math Competition Book Division E 2016-2017
ZIML Math Competition Book Division M 2016-2017
ZIML Math Competition Book Division H 2016-2017
ZIML Math Competition Book Jr Varsity 2016-2017
ZIML Math Competition Book Varsity Division 2016-2017
ZIML Math Competition Book Division E 2017-2018
ZIML Math Competition Book Division M 2017-2018
ZIML Math Competition Book Division H 2017-2018
ZIML Math Competition Book Jr Varsity 2017-2018
ZIML Math Competition Book Varsity Division 2017-2018

MATH CHALLENGE CURRICULUM TEXTBOOKS SERIES

Math Challenge I-A Pre-Algebra and Word Problems
Math Challenge I-B Pre-Algebra and Word Problems
Math Challenge I-C Algebra
Math Challenge II-A Algebra
Math Challenge II-B Algebra
Math Challenge III Algebra
Math Challenge I-A Geometry
Math Challenge I-B Geometry
Math Challenge I-C Topics in Algebra
Math Challenge II-A Geometry
Math Challenge II-B Geometry

Math Challenge III Geometry
Math Challenge I-B Counting and Probability
Math Challenge II-A Combinatorics
Math Challenge I-B Number Theory
Math Challenge II-A Number Theory

COMING SOON FROM ARETEEM PRESS

Fun Math Problem Solving For Elementary School Vol. 2 (and Solutions Manual)
Counting & Probability for Middle School (and Solutions Manual) - From Common Core to Math Competitions
Number Theory Problem Solving for Middle School (and Solutions Manual) - From Common Core to Math Competitions
Other volumes in the **Math Challenge Curriculum Textbooks Series**

The books are available in paperback and eBook formats (including Kindle and other formats).

To order the books, visit https://areteem.org/bookstore.

Contents

Introduction

Each month during the school year, Areteem Institute hosts the online Zoom International Math League (ZIML) competitions. Students can compete in one of five divisions based on their age and mathematical level (details shown on Page 9).

This book contains the problems, answers, and full solutions from the nine ZIML Division M Competitions held during the 2017-2018 School Year. It is divided into three parts:

1. The complete Division M ZIML Competitions (20 questions per competition) from October 2017 to June 2018.
2. The solutions for each of the competitions, including detailed work and helpful tricks.
3. An appendix including the topics and knowledge points covered for Division M, a glossary including common mathematical terms, and answer keys for each of the competitions so students can easily check their work.

The questions found on the ZIML competitions are meant to test your problem solving skills and train you to apply the knowledge you know to many different applications. We hope you enjoy the problems!

About Zoom International Math League

The Zoom International Math League (ZIML) has a simple goal: provide a platform for students to build and share their passion for math and other STEM fields with students from around the globe. Started in 2008 as the Southern California Mathematical Olympiad, ZIML has a rich history of past participants who have advanced to top tier colleges and prestigious math competitions, including American Math Competitions, MATHCOUNTS, and the International Math Olympaid.

The ZIML Core Online Programs, most available with a free account at ziml.areteem.org, include:

- **Daily Magic Spells:** Provides a problem a day (Monday through Friday) for students to practice, with full solutions available the next day.
- **Weekly Brain Potions:** Provides one problem per week posted in the online discussion forum at ziml.areteem.org. Usually the problem does not have a simple answer, and students can join the discussion to share their thoughts regarding the scenarios described in the problem, explore the math concepts behind the problem, give solutions, and also ask further questions.
- **Monthly Contests:** The ZIML Monthly Contests are held the first weekend of each month during the school year (October through June). Students can compete in one of 5 divisions to test their knowledge and determine their strengths and weaknesses, with winners announced after the competition.
- **Math Competition Practice:** The Practice page contains sample ZIML contests and an archive of AMC-series tests for online practice. The practices simulate the real contest environment with time-limits of the contests automatically controlled by the server.
- **Online Discussion Forum:** The Online Discussion Forum

is open for any comments and questions. Other discussions, such as hard Daily Magic Spells or the Weekly Brain Potions are also posted here.

These programs encourage students to participate consistently, so they can track their progress and improvement each year.

In addition to the online programs, ZIML also hosts onsite Local Tournaments and Workshops in various locations in the United States. Each summer, there are onsite ZIML Competitions at held at Areteem Summer Programs, including the National ZIML Convention, which is a two day convention with one day of workshops and one day of competition.

ZIML Monthly Contests are organized into five divisions ranging from upper elementary school to advanced material based on high school math.

- **Varsity:** This is the top division. It covers material on the level of the last 10 questions on the AMC 12 and AIME level. This division is open to all age levels.
- **Junior Varsity:** This is the second highest competition division. It covers material at the AMC 10/12 level and State and National MathCounts level. This division is open to all age levels.
- **Division H:** This division focuses on material from a standard high school curriculum. It covers topics up to and including pre-calculus. This division will serve as excellent practice for students preparing for the math portions of the SAT or ACT. This division is open to all age levels.
- **Division M:** This division focuses on problem solving using math concepts from a standard middle school math curriculum. It covers material at the level of AMC 8 and School or Chapter MathCounts. This division is open to all students who have not started grade 9.

- **Division E:** This division focuses on advanced problem solving with mathematical concepts from upper elementary school. It covers material at a level comparable to MOEMS Division E. This division is open to all students who have not started grade 6.

This problem book features the Division M Contests. For a detailed list of topics covered for Division M see p.173 in the Appendix.

About Areteem Institute

Areteem Institute is an educational institution that develops and provides in-depth and advanced math and science programs for K-12 (Elementary School, Middle School, and High School) students and teachers. Areteem programs are accredited supplementary programs by the Western Association of Schools and Colleges (WASC). Students may attend the Areteem Institute through these options:

- Live and real-time face-to-face online classes with audio, video, interactive online whiteboard, and text chatting capabilities;
- Self-paced classes by watching the recordings of the live classes;
- Short video courses for trending math, science, technology, engineering, English, and social studies topics;
- Summer Intensive Camps on prestigious university campuses and Winter Boot Camps;
- Practice with selected daily problems for free, and monthly ZIML competitions at ziml.areteem.org.

The Areteem courses are designed and developed by educational experts and industry professionals to bring real world applications into STEM education. The programs are ideal for students who wish to build their mathematical strength in order to excel academically and eventually win in Math Competitions (AMC, AIME, USAMO, IMO, ARML, MathCounts, Math Olympiad, ZIML, and other math leagues and tournaments, etc.), Science Fairs (County Science Fairs, State Science Fairs, national programs like Intel Science and Engineering Fair, etc.) and Science Olympiad, or purely want to enrich their academic lives by taking more challenges and developing outstanding analytical, logical thinking and creative problem solving skills.

Since 2004 Areteem Institute has been teaching with methodology that is highly promoted by the new Common Core State Standards: stressing the conceptual level understanding of the math concepts, problem solving techniques, and solving problems with real world applications. With the guidance from experienced and passionate professors, students are motivated to explore concepts deeper by identifying an interesting problem, researching it, analyzing it, and using a critical thinking approach to come up with multiple solutions.

Thousands of math students who have been trained at Areteem achieved top honors and earned top awards in major national and international math competitions, including Gold Medalists in the International Math Olympiad (IMO), top winners and qualifiers at the USA Math Olympiad (USAMO/JMO), and AIME, top winners at the Zoom International Math League (ZIML), and top winners at the MathCounts National. Many Areteem Alumni have graduated from high school and gone on to enter their dream colleges such as MIT, Cal Tech, Harvard, Stanford, Yale, Princeton, U Penn, Harvey Mudd College, UC Berkeley, UCLA, etc. Those who have graduated from colleges are now playing important roles in their fields of endeavor.

Further information about Areteem Institute, as well as updates and errata of this book, can be found online at http://www.areteem.org.

Acknowledgments

This book contains the Online ZIML Division M Problems from the 2017-18 school year. These problems were created and compiled by the staff of Areteem Institute. These problems were inspired by questions from the Areteem Math Challenge Courses, past questions on the ACT/SAT/GRE, past math competitions, math textbooks, and countless other resources and people encountered by the Areteem Curriculum Department in their life devoted to math. We thank all these sources for growing and nurturing our passion for math.

The Areteem staff, including John Lensmire, David Reynoso, Kevin Wang, and Kelly Ren, are the main contributors who compiled, edited, and reviewed this book.

Lastly, thanks to all the students who have participated and continue to participate in the Zoom International Math League. Your dedication to the Daily Magic Spells and Monthly Contests makes all of this possible, and we hope you continue to enjoy ZIML for years to come!

1. ZIML Contests

This part of the book contains the Division M ZIML Contests from the 2017-18 School Year. There were nine monthly competitions, held on the dates found below:

- October 6-8
- November 3-5
- December 1-3
- January 5-7
- February 2-4
- March 2-4
- April 6-8
- May 4-6
- June 1-3

1.1 ZIML October 2017 Division M

Below are the 20 Problems from the Division M ZIML Competition held in October 2017.

The answer key is available on p.184 in the Appendix.

Full solutions to these questions are available starting on p.92.

Problem 1
If 5 ants eat one whole cube of sugar in 3 minutes, how many minutes would 2 ants need to eat 4 cubes of sugar?

Problem 2
If Rachel and Anna work together, they can finish a project in 90 hours. Working separately, it takes the same time for Rachel to complete $\frac{1}{5}$ of the project as it takes Anna to complete $\frac{1}{3}$ of the project. If Anna works alone on the project, how many hours would she need to work to complete it?

Problem 3

A rectangle is divided into 6 squares, as shown in the diagram.

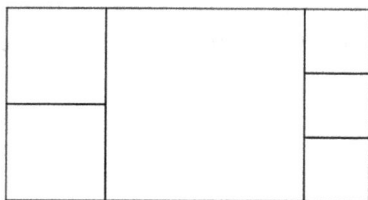

The area of the smallest square is 32 less than the area of the big square. What is the area of the whole rectangle?

Problem 4

You have been keeping record of the time it takes you to complete a 100 m sprint. So far have run 7 sprints with an average time of 14.9 seconds. If you want to improve your average to 14.8 seconds after the next attempt, how fast, in seconds, does your next attempt need to be?

Problem 5

A 3-digit number leaves a remainder of 4 when divided by 5, and a remainder of 5 when divided by 6. What is the largest number with these properties?

Problem 6

Harriet is working on an assignment for her English class. She had to answer several questions about a Shakespeare novel. So far, the questions she has answered and the questions she has not answered are in ratio 4 : 5. After she answers 5 more questions, the ratio becomes 5 : 4. How many questions in total are on the assignment?

Problem 7

You flip a coin 3 times. The probability that you get more tails than heads is $\dfrac{P}{Q}$ as a reduced fraction. What is $Q - P$?

Problem 8

There are 40 cars in a parking lot, all either red or blue. Further, some are electric while the rest use gas. 20 of the cars are red and 15 of the cars are electric. If 12 of the cars are blue and use gas, how many red electric cars are in the parking lot?

Problem 9

In the following diagram, *ABCD* is a parallelogram, and *F* is the midpoint of *DC* and *AD* = *DE*.

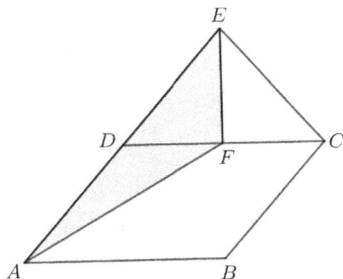

If the area of parallelogram *ABCD* is 64, what is the area of the shaded region?

Problem 10

You and 6 of your friends just formed a Math Zoom Club. In how many ways can you choose a president and a vice-president?

Problem 11

A pool has a pump that delivers hot water and a pump that delivers cold water. It takes 17 hours to fill the pool with only cold water and 34 hours to fill the pool with only hot water. The maintenance guy opened the pump of cold water but forgot about the pump of hot water. After 5 hours he came back and opened the pump of hot water as well, and both pumps stayed open until the pool was full. In total, how many hours was the pump of cold water open?

Problem 12

Consider the "star" diagram below (the diagram is not drawn to scale).

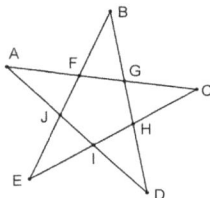

If $\angle EHB = \angle EFC = 105°$, and $\triangle GHC$ is isosceles, what is the measure of $\angle E$ in degrees?

Problem 13

Madam Lyndy's restaurant offers 3 choices of appetizers, 10 choices of entrees, and 4 choices of dessert. Jake and Jacky want to order 1 appetizer to share, and each wants their own entree and their own dessert. They want to order different desserts so they can try each other's. In how many different ways can they choose their dinner?

Problem 14

A number $\overline{a25b0}$ is divisible by 6 and 8, and has no repeated digits. What is the number?

Problem 15

What is the units digit of 2017^{2017}?

Problem 16
How many triangles are there in the following diagram?

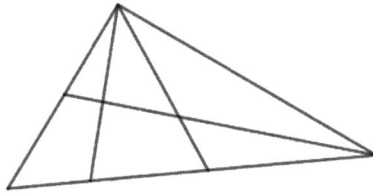

Problem 17
The least common multiple of two numbers is 2520 and their greatest common divisor is 20. If one of the numbers is 40, what is the other number?

Problem 18
The area of a regular hexagon is $54\sqrt{3}$. What is the side length of the hexagon?

Problem 19
The weekend before Halloween, 2 out of every 3 people that entered the local grocery store left with a bag of apples, and 1 out of every 7 people that entered the store left with a bag of candy corn. Tony was supposed to count how many people entered the store, but he is unsure of the exact number. He does remember the number was bigger than 120 and no more than 144. How many people bought candy corn for Halloween over the weekend?

Problem 20

Rick's dad just bought some old books from a discount store and asked Rick for help deciding how to arrange them on a shelf. He got 4 math books, 3 physics books, and 2 biology books. He wants to make sure that all physics books are placed together (the other subjects do not need to be together). In how many different ways can Rick arrange the books on the shelf?

1.2 ZIML November 2017 Division M

Below are the 20 Problems from the Division M ZIML Competition held in November 2017.
The answer key is available on p.185 in the Appendix.
Full solutions to these questions are available starting on p.99.

Problem 1
Two squares with side length 21 are attached together as shown in the diagram below.

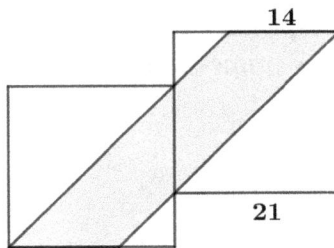

Find the area of the shaded region.

Problem 2
Rita has 36 marbles, 20 of which are red and 16 of which are white. Rose has 27 marbles, all of them either red or white. Suppose Rita and Rose have the same ratio of red to white marbles. How many white marbles does Rose have?

Problem 3

Jane's family loves to come visit her. Her parents come visit every 6 weeks, and her grandparents come visit every 8 weeks. Both Jane's parents and grandparents visited during the same week the last week of 2016. How many times will her parents and grandparents visit during the same week in 2017? (Remember one year has 52 weeks.)

Problem 4

Call a prime number that yields a prime number when its units digit is removed a "prime-prime number". For example, 131 is a three-digit prime-prime number because 131 is prime and 13 is prime. How many prime-prime numbers are between 200 and 250?

Problem 5

Mark has 21 identical pieces of candy saved to bring to lunch next week. He wants to make sure to bring at least 3 pieces each day, except for Friday when he wants to bring at least 5. In how many ways can he sort his candy to bring each of the 5 days of the week to lunch?

Problem 6

A math class with 20 students take an exam. A fourth of the students who passed got an A and a third who passed got a B. How many students passed the exam?

Problem 7

Jack's brother can clean the kitchen in 60 minutes. If Jack helps his brother, they can clean it in 45 minutes. How many minutes would it take Jack to clean it himself?

Problem 8

Suppose in the diagram below that $\triangle ABC$ is isosceles and $\angle FEB = 165°$. Find the measure of $\angle GAB$ in degrees.

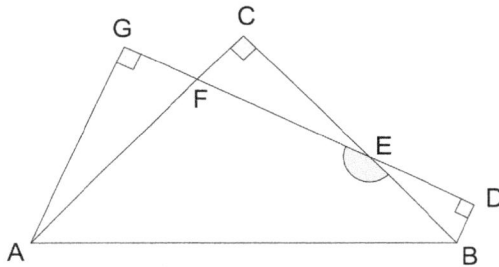

Problem 9

Three squares with side lengths 4, 8 and 4 are arranged as in the diagram below.

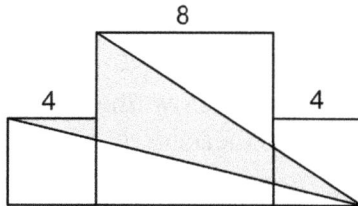

What is the area of the shaded region?

Problem 10

Michael offered Jim a bonus for his weekly sales. Jim would get a 6% bonus on the first $500 of weekly sales and 10% bonus on any sales past $500. In addition, if Jim managed to sell at least $2500 in one week, that entire week's bonus would be increased by 20%. If Jim sold $2000 and $3200 the following two weeks, what would his bonus be for those two weeks? Round your answer to the nearest dollar if necessary.

Problem 11

How many numbers between 140 and 700 have an odd number of factors?

Problem 12
How many positive 2-digit numbers are factors of both 3600 and 7560?

Problem 13
At 6 AM, bus station A starts to dispatch buses to station B, and station B starts to dispatch buses to station A. They each dispatch one bus to the other station every 12 minutes. The one-way trip takes 52 minutes. One passenger gets on the bus at station A at 6:24 AM. How many buses coming from station B will the passenger see en route?

Problem 14
Trapezoid $ABCD$ is such that $AB = 12$, $CD = 24$ and its height is 8.

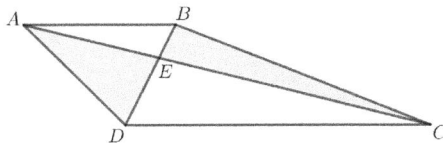

If the area of $\triangle EDC$ is 64, written $[EDC] = 64$, what is the area of the shaded region?

Problem 15
How many rearrangements of the letters in the word "HAL-LOWEEN" are there?

Problem 16

A number $\overline{32a5b}$ is divisible by 11. Its last two digits form a number $\overline{5b}$ that can be exactly divided by 3. What is the largest 5-digit number with this with these properties?

Problem 17

Consider the shaded region in the diagram below, formed from a right triangle and a sector of a circle.

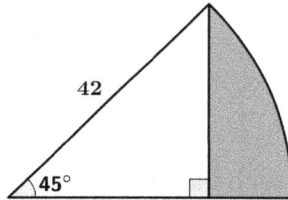

Using the approximation $\pi = \dfrac{22}{7}$, the area of the shaded region is an integer L. What is L?

Problem 18

What is the largest 3-digit number that leaves a reminder of 7 when divided by 8, 10 and 12?

Problem 19

Suppose you pay $5 to play a game at the carnival. The game is simple, you press a button and a dial randomly picks a number on a number line between 2 and 14.

2.71

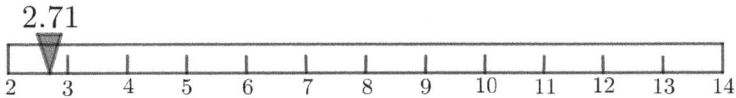

You then win the amount of money shown (in dollars). What percentage of the time will you win at least as much as you paid to play the game? Input your answer rounded to the nearest percent. For example 82.5% should be submitted as 83.

Problem 20

How many gallons of a 10% ammonia solution should be mixed with 50 gallons of a 30% ammonia solution to make a 15% ammonia solution?

1.3 ZIML December 2017 Division M

Below are the 20 Problems from the Division M ZIML Competition held in December 2017.
The answer key is available on p.186 in the Appendix.
Full solutions to these questions are available starting on p.107.

Problem 1
Grace begins walking at a pace of 5 km per hour from one end of the trail that is 60 km long. Danny begins at the other end of the trail at the same time, walking towards Grace at a pace of 7 km per hour. How long will it take for them to pass each other in hours?

Problem 2
A machine randomly chooses numbers between 1 and 10 (inclusive). So far the machine has chosen the numbers 3, 7, 2, 9, 4, and 4. What number should come next so that the average of all the numbers is 5?

Problem 3

Arrange several equilateral triangles and rhombi, all of whose side lengths are 3 cm, to form a long parallelogram, as shown in the diagram.

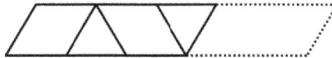

If the perimeter of the long parallelogram is 438 cm, how many rhombi are there?

Problem 4

A group of 68 people rent 24 motorcycles of two kinds at a racetrack. The first kind has a capacity of 2 people and costs $40 per motorcycle. The second has a capacity of 3 people and costs $30 per motorcycle. The 68 people exactly fill all vehicles. What is the total cost in renting the 24 motorcycles in dollars?

Problem 5

The sum of eleven consecutive even numbers is 616. What is the second smallest number?

Problem 6

Jong-Zhi took a math test that had 12 arithmetic questions, 15 algebra questions and 18 geometry questions. She got 75% of the arithmetic questions correct and 60% of the algebra questions correct. How many of the geometry questions must she get correct to get at least a passing grade of 75%?

Problem 7

Suppose we are given a bag of 30 balls and 20 cubes. It is known that of the 50 objects in the bag, 40 of them are red and 10 of them are blue. If at least one of the objects in the bag is a blue cube, what is the minimum number of red balls in the bag?

Problem 8

What is the next term in the sequence $2, 3, 10, 29, 66, 127, \_\_$?

Problem 9

In the diagram below, there are 21 grid points arranged in equilateral triangles, equally spaced. The AREA of each small equilateral triangle formed by 3 adjacent grid points is 1. Find the area of $\triangle ABC$.

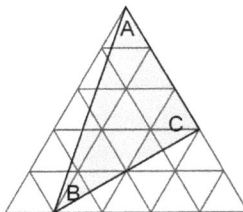

Problem 10

How many six-digit numbers represented by 2017__ __ can be exactly divided by 3 and 5?

Problem 11

How many ways can you place 20 identical balls into 4 distinguishable bins such that there is at least one ball in each bin?

Problem 12

All the smaller circles in the diagram have radii 1. The shaded region has perimeter $K\pi$ for some integer K. What is K?

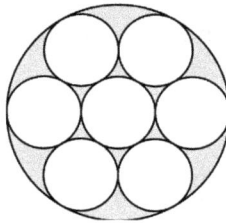

Problem 13

What is the smallest number K such that $702 \times K$ is a perfect square?

Problem 14

Find the distance from one corner to the opposite corner (for example the lower front left vertex to the upper back right vertex) of a rectangular prism (a box) with dimensions $3, 4, 12$.

Problem 15

Katie went hiking on a hill near her home. From the bottom of the hill, She went up to the top and then came down along the same trail, back to the spot she started. Assume her uphill speed was 4 miles per hour, and her downhill speed was 6 miles per hour. What is her average speed for the whole uphill-downhill trip in miles per hour? Round your answer to the nearest tenth if necessary.

Problem 16

Suppose you have 5 red cards and 5 black cards. You randomly are dealt 5 of the cards. The probability you get 3 red cards and 2 black cards is $\dfrac{P}{Q}$ as a reduced fraction. What is $Q - P$?

Problem 17

In the diagram below the small circle has radius 1 and the big circle has radius 2.

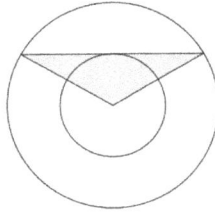

What is the area of the shaded region? Round your answer to the nearest tenth if necessary.

Problem 18

The number of Harry Potter fans and Twilight fans that attended last year's comic-con are in ratio 5 : 4. The number of Start Trek fans and Harry Potter fans that attended the comic-con are in ratio 5 : 3. If there were 600 more Harry Potter fans than Twilight fans at the comic-con, how many Star Trek fans attended?

Problem 19

What is the remainder of $7^{2017} + 5^{2017}$ when we divide the sum by 6?

Problem 20

You and 6 of your friends are camping in the beach over the weekend (staying for two nights). For each of the two nights you agree to have two shifts of 2 people keeping watch while the others sleep. To make it fair each night, if someone keeps watch on the first shift of a night, that person will not keep watch on the second shift. In how many different ways can you decide the 4 shifts of keeping watch?

1.4   ZIML January 2018 Division M

Below are the 20 Problems from the Division M ZIML Competition held in January 2018.
The answer key is available on p.187 in the Appendix.
Full solutions to these questions are available starting on p.116.

Problem 1
Dakota got 85% of the questions correct on a 20 question test, and 70% of the questions correct on a 30 question test. What percent of the questions did he get correct for both tests combined?

Problem 2
Consider the trapezoid $ABCD$ with AB parallel to CD. Let E be the intersection of the diagonals AC and BD, and suppose $AB = 10$, $CD = 15$, and $\triangle ABE$ has area 36. What is the area of triangle CDE?

Problem 3
How many positive integers are factors of both 160 and 60?

Problem 4
Stephanie begins walking at a pace of 4 km per hour from one end of the trail that is 34 km long. Bob begins at the other end of the trail at the same time, walking towards Stephanie at a pace of 6 km. How many hours will it take for them to pass each other? Round your answer to the nearest tenth if necessary.

Problem 5

If you randomly pick an integer number between 54 and 72, inclusive, the probability that the number is a multiple of 7 is $\frac{P}{Q}$ in lowest terms. What is $Q - P$?

Problem 6

In the following diagram $ABCDEF$ is a regular hexagon with area 114, and O is its center.

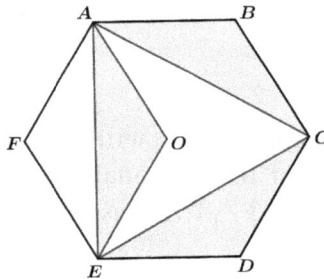

What is the area of the shaded region?

Problem 7

There are 3 pupils whose ages are three consecutive integers. The product of their ages is 720. What is the sum of their ages?

Problem 8

There are 4 blue cards and 5 red cards. All cards are shuffled and you are dealt 2 of them. The probability that the cards are of different color is $\dfrac{P}{Q}$ in lowest terms. What is $Q - P$?

Problem 9

In the diagram below the squares have side lengths 1, 2 and 3.

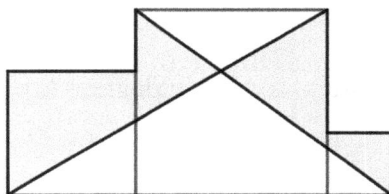

What is the area of the shaded region? Express your answer as a decimal rounded to the nearest hundredth if necessary.

Problem 10

A 4-digit number consists of different digits and is divisible by 3, 5 and 7. What is the largest number that satisfies these conditions?

Problem 11

In the following diagram, $\angle BAC = 47°$, $\angle CDE = 24°$, $\angle EFG = 85°$, and $\angle EGF = 15°$.

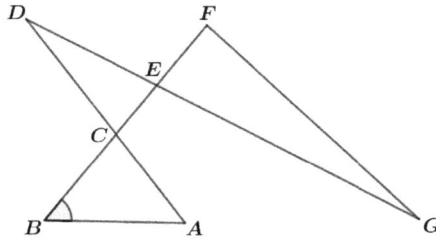

What is the measure of $\angle ABC$ in degrees?

Problem 12

Peter painted $\frac{1}{3}$ of a room while Richard painted $\frac{2}{5}$ of the same room. It then took Peter 1 hour, 40 minutes to finish painting the remainder of the room by himself. If Peter had painted the entire room by himself, how many hours would it have taken? Express your answer in hours, rounded to the nearest hundredth if necessary.

Problem 13

A 2-digit integer leaves a remainder of 5 when it is divided by 6 and a remainder of 4 when it is divided by 5. What is the biggest number that satisfies these conditions?

Problem 14

What is the smallest 6-digit number with no repeated digits that starts with 9 and is a multiple of 11?

Problem 15

A spider has 8 legs. A firefly has 6 legs and 2 pairs of wings. A cicada has 6 legs and 1 pair of wings. There are a total of 21 bugs of the three types. There are 138 legs in total. There are 23 pairs of wings in total. How many fireflies are there?

Problem 16

There are 10 identical pigeons that will occupy 4 numbered pigeonholes at a park overnight. The pigeonholes are not too big, so they can each fit at most 3 pigeons. In how many different ways can the pigeons occupy the 4 pigeonholes?

Problem 17

The hexagon in the diagram below is made up of small equilateral triangles. 22 of these equilateral triangles are shaded, as shown below.

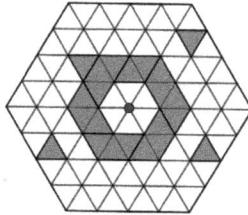

What is the minimum number of additional triangles that should be shaded so that if the figure is rotated 60°, 120°, 180°, 240°, or 300° around its center, the figure looks the same before and after rotation?

Problem 18

There are 3 one-way paths that go from city A to city B, 1 one-way path from city B to city A, 4 two-way paths that connect city B and city C, 2 one-way paths from city A to city C and 3 one-way paths from city C to city A. How many loops are there from city A back to city A without visiting any other city twice?

Problem 19

In the warehouse there were 3 times as many pounds of apples as bananas at the beginning. Suppose 250 pounds of bananas and 600 pounds of apples were sold every day, and a few days later the bananas were sold out and 750 pounds of apples were left. How many pounds of apples were there originally?

Problem 20

You have three 6-sided dice, colored blue, green, and red. If you throw all three dice and add up the numbers shown, how many different outcomes add up to 5?

1.5 ZIML February 2018 Division M

Below are the 20 Problems from the Division M ZIML Competition held in February 2018.
The answer key is available on p.188 in the Appendix.
Full solutions to these questions are available starting on p.126.

Problem 1
What is the largest 2-digit number that has an odd number of factors?

Problem 2
Dev was buying some clothes on sale at a department store. The sign stated that each item had a 10% discount and then an extra 10% discount on top of that. Apparently the cashier made a mistake, and gave Dev a 20% discount instead. When he realized the mistake, Dev went back to the store to give back the difference in price so the cashier wouldn't get in trouble. If Dev had paid $160, how many dollars will he give back to the cashier? Round your answer to the nearest cent if necessary (that is, to the nearest hundredth in dollars).

Problem 3
What is the largest 5-digit number that can be constructed using each of the digits 1, 2, 3, 4 and 5, such that the sum of every pair of consecutive digits is a prime number?

Problem 4

Consider a square and a rectangle, both with area 16. Suppose that the length of the rectangle is twice the side length of the square. The ratio of the perimeter of the square to the perimeter of the rectangle is $a : b$, where a and b have no common factors. What is $a + b$?

Problem 5

Larry and Moe are on opposite sides of a 1.5 mile long track. Larry starts running first towards Moe, and some time later Moe starts running towards Larry. They both run at the same constant speed of 10 mph, and stop running when they meet on the track. If Larry runs for double the time as Moe, how many minutes does Larry run?

Problem 6

Kelsey has a deck of 15 cards with numbers 1, 2, 3, 4 and 5 (three of each). If she chooses 5 cards at random the probability that she gets 3 cards with the same number and a pair of cards with the same number is $\dfrac{P}{Q}$ in lowest terms. What is $Q - P$?

Problem 7

A square is divided into 7 congruent rectangles as shown on the diagram below.

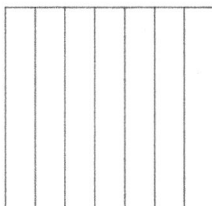

If the area of 3 rectangles combined is 84, what is the width of one of the rectangles?

Problem 8

A number between 40 and 90 leaves a remainder of 3 when divided by 7, and a remainder of 7 when divided by 13. What is the number?

Problem 9

Tom drives his car for a round trip between place A and place B. He drives at 40 km per hour to get from A to B. At what speed should he drive back from B to A, if his average speed for the round trip is 48 km per hour? Give your answer in km per hour and round to the nearest tenth if necessary.

Problem 10
At the ice cream shop you can customize your ice cream by choosing if you want (i) regular cone or waffle cone, (ii) 1 flavor of ice cream or 2 flavors of ice cream mixed together, and (iii) sprinkles or no sprinkles. If there are 3 ice cream flavors available to choose from, how many different ways can you customize your ice cream?

Problem 11
In the diagram below, $ABCD$ is a parallelogram and E is a point on the diagonal AC such that $EC = 2AE$.

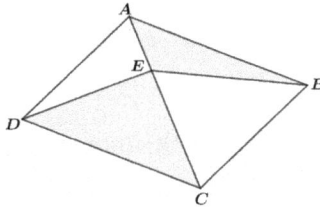

If the shaded area is 24, what is the area of the parallelogram $ABCD$?

Problem 12
Liam had 7 tiles with the letters $\boxed{C}\,\boxed{A}\,\boxed{B}\,\boxed{B}\,\boxed{A}\,\boxed{G}\,\boxed{E}$. He put all the tiles in a bag and then took them out one by one to form a 7 letter word. The probability that he formed the word CABBAGE is $\dfrac{P}{Q}$ in lowest terms. What is $P+Q$?

Problem 13

Carlos always leaves his cell phone on. If his cell phone is on and he is playing the mobile game Pokemon Go, it will last for only 4 hours. However, if it is on and he is not playing Pokemon Go it will last for 18 hours.

Since the last recharge, his phone has been on 5 hours, and during that time he has been playing Pokemon Go for 2 hours. If he stops playing Pokemon Go (but leaves the phone on), how many more hours will the battery last?

Problem 14

You roll three identical 6-sided dice and record how many of each number you get, but not their order. For example, you could record 1 three and 2 fives or you could record 3 sixes. How many different outcomes are there?

Problem 15

Consider square $ABCD$ and two points E and F outside the square such that ABE and BCF are equilateral triangles. What is the angle measure in degrees of $\angle BEF$?

Problem 16

Find the product of all the factors of 24.

Problem 17

A candy shop sold three flavors of candies, cherry, strawberry, and watermelon, in the morning. The prices are $20/kg, $25/kg, and $30/kg, respectively. The shop sold a total of 100 kg and received $2570. It is known that the total sale of cherry and watermelon flavor candies combined is $1970. How many kilograms of watermelon flavor candies were sold?

Problem 18

Roy just learned in geometry class that the lengths of the sides of a triangle are always such that the sum of the lengths of any two of them is always larger than the length of the remaining side. If Roy has sticks of lengths 2 inches, 3 inches, 4 inches, 5 inches, and 6 inches (one of each), how many different triangles can Roy make with 3 of the sticks?

Problem 19

What is the units digit of $35^{33} + 53^{33}$?

Problem 20

Your family is having dinner together with some friends around a circular table. Your parents want to sit next to each other, and 2 friends that came together also want to sit next to each other. You and the remaining 3 friends do not care where you sit. How many different ways can the 8 people choose where to sit? (The specific chairs chosen do not matter.)

1.6 ZIML March 2018 Division M

Below are the 20 Problems from the Division M ZIML Competition held in March 2018.
The answer key is available on p.189 in the Appendix.
Full solutions to these questions are available starting on p.134.

Problem 1

Each student in Ms. Nixon's class randomly picked an integer between 1 and 10 (inclusive). They came up with the following list of numbers:

$$4, 6, 10, 3, 7, 8, 2, 5, 1, 10, 9, 5, 7.$$

Percival was late to class, so when he arrived he was asked by Ms. Nixon to also choose a number and add it to the list. The probability that Percival's number will not increase the median of the numbers in the list is $\frac{P}{Q}$ in lowest terms. What is $Q - P$?

Problem 2

In the following diagram the circle has area π and the vertices of
the small square are in the midpoints of the sides of the bigger
square.

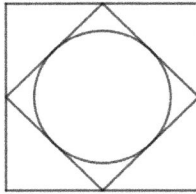

What is the area of the bigger square? Round your answer to the
nearest tenth if necessary.

Problem 3

Among the fractions

$$\frac{1}{30}, \frac{2}{30}, \frac{3}{30}, \ldots, \frac{29}{30},$$

how many are irreducible? (Irreducible means the fraction cannot
be simplified.)

Problem 4

The table below shows the percent of students that study Math and Computer Science at the University of Honolulu.

Year	1st	2nd	3rd	4th
Math	32%	29%	24%	15%
C.S.	29%	25%	24%	22%

If there are 200 Math students and 500 Computer Science students, what percent of the students are in their 4th year at school? Round your answer to the nearest percent if necessary.

Problem 5

In the following diagram the pentagon has area 56 and each of the circles has area 2.

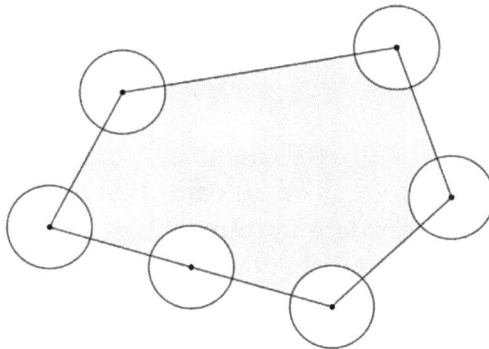

What is the shaded area?

Problem 6

Edward and Sydney have some jellybeans in their hands. Edward has 2 green, 1 red, and 3 blue. Sydney has 3 red, 1 green, and 5 yellow. If each of them randomly picks one of their jellybeans, the probability that they pick jellybeans of different colors is $\dfrac{P}{Q}$ in lowest terms. What is $P + Q$?

Problem 7

Dane and Crystal want to see each other so they decide to drive from their house towards each other using the same road and meet somewhere in the middle. Both of them drive at a constant speed during the entire trip. At the same speed they are driving towards each other, it would take Dane 180 minutes to reach Crystal's house, and it would take Crystal 120 minutes to reach Dane's house. If they leave at the same time, after how many minutes would they meet on the road?

Problem 8

There is a stack with 2018 numbered cards (from 1 to 2018). Corinne decided to remove all cards that ended in 0, and then number them again starting from 1. If she removes once more the cards that have a number that ends in 0, how many cards are left?

Problem 9

Quintin lives 5 blocks south and 4 blocks east of his friend Renee. They live in a city where all blocks are the same size and form a square grid. Quintin recently learned that there is a new ice cream shop 3 blocks north and 2 blocks west of his house. Quintin wants to visit Renee taking the shortest path possible and making sure he visits the ice cream so he can bring some joy to his friend. In how many different ways can Quintin plan his route?

Problem 10

In the following diagram, there are 28 grid points arranged in equilateral triangles, equally spaced. The area of each small triangle formed by 3 adjacent grid points is 1.

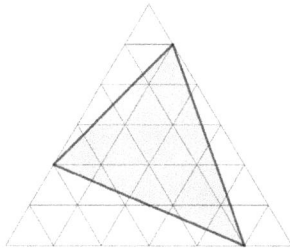

What is the area of the shaded region? Round your answer to the nearest tenth if necessary.

Problem 11

How many factors of 2160 are perfect cubes?

Problem 12
A box of dark chocolate costs \$3, a box of milk chocolate costs \$2, and a box of white chocolate costs \$4. Wallace bought 8 boxes of chocolate and spent \$23 in total. At most how many of the boxes could have been of white chocolate?

Problem 13
Every one and a half cats eat a mouse and a half in one and a half hours. How many mice can 15 cats eat in 15 hours?

Problem 14
What 3-digit number would result in the largest number of trailing zeros when we multiply it by 320? (For example $320 \times 100 = 32000$ results in 3 trailing zeros.)

Problem 15

Flo was busy drawing circles in a piece of paper. All of her circles were identical and sometimes they overlapped. She noticed that when two circles overlapped the resulting figure was split in three regions, and when three circles overlapped, the resulting figure was split into at most 7 regions, as shown in the diagram below.

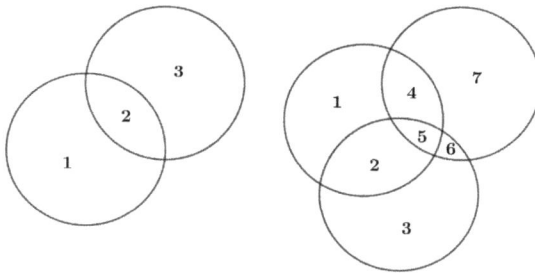

If she draws 4 identical circles that overlap with each other, what is the largest number of regions she can split her figure into?

Problem 16

Demetria has a bag with socks. The bag has 12 pairs of sport socks and 18 pairs of dress socks. Each pair of socks is either purple or green. Demetria is sure that she has at least one pair of purple dress socks, but does not remember how many of each of the other kinds of pairs she has. If there are 19 pairs of green socks and 11 pairs of purple socks, what is the minimum possible number of pairs of green sport socks in the bag?

Problem 17

Dary, the explorer ant, needed to explore an area in the shape of an equilateral triangle of side length 200 feet. She started at one of the vertices and followed a zig-zag like path. Part of the path she followed is shown in the diagram below.

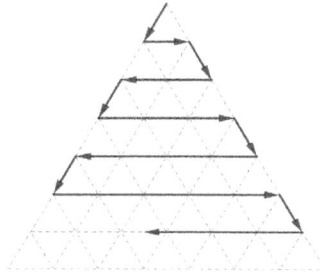

Each of the small triangles in the imaginary grid that Dary used to guide herself has side length 10 feet. If she stopped exploring when she visited all three vertices of the big triangle, how many feet did she walk?

Problem 18

There is a line with 12 lizards in the middle of the desert. As part of their spring solstice festival, the lizards perform a dance at noon on March 21st. During the first 12 minutes after 12:00, the lizards dance as follows: (i) at the end of the first minute all lizards flip over (so all of their bellies are facing up); (ii) at the end of the second minute all the lizards in an even position flip over (so all odd lizards still have their bellies facing up, but all even lizards have their backs facing up); (iii) at the end of the third minute all lizards in a position that is a multiple of 3 flip over (so if they their bellies were facing up now their backs are facing up and vice versa). They continue dancing to this pattern until 12 minutes have passed. At the end of the dance, how many lizards have their bellies facing up?

Problem 19

A water tank is missing 40% of water. It has 40 more gallons of water than when it is 40% full. How many gallons of water does the tank have when it is full?

Problem 20

The corners of an 8×8 checker board were removed, producing a board like the one in the diagram below.

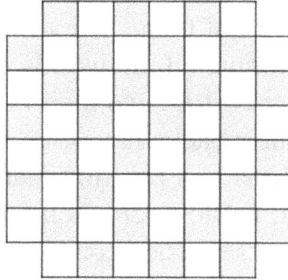

In how many ways can a black checker piece and a white checker piece be placed on the board so that they are adjacent to each other? (Two checker pieces will be adjacent to each other if they are within one square of each other horizontally, vertically or diagonally.)

1.7 ZIML April 2018 Division M

Below are the 20 Problems from the Division M ZIML Competition held in April 2018.
The answer key is available on p.190 in the Appendix.
Full solutions to these questions are available starting on p.144.

Problem 1
In the following diagram 6 circles are arranged around the inside of a larger circle.

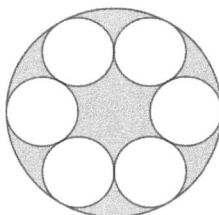

All the small circles have radius 2. What is the area of the shaded region? Round your answer to the nearest integer if necessary.

Problem 2
When Dixie distributed her candies evenly among 5 of her friends and herself she had 3 candies left over. The next day she started with the same number of candies as the day before, she ate 3 before distributing the rest evenly among her 5 friends, and had 4 left over. What is the least number of candy Dixie could have had at the beginning of each day?

Problem 3

Brock has two cookie jars. One is in the shape of Cookie Monster, and the other one is in the shape of Elmo. As the Cookie Monster loves cookies way more than Elmo, she makes sure to have way more cookies in the Cookie Monster jar than in the Elmo jar. She just baked a batch of cookies and she put 96 of them inside of the Cookie Monster jar. If the number of cookies she put inside the Cookie Monster and Elmo jars are in ratio 8 : 3, how many cookies did she bake in total?

Problem 4

Raul bought a lottery ticket where he must choose 5 different numbers from 1 to 15 as well as 1 of 5 different symbols. In how many ways could Raul buy his lottery ticket?

Problem 5

Mrs. Gatsby has tons of cats and yellow canaries at home, 116 of them in total. If there are 58 more yellow canary legs than cat legs in her house, how many yellow canaries does she have? (Each cat has 4 legs and each canary 2 legs.)

Problem 6

In an expedition to the local park, all the students in Matty's class were asked to bring a canteen for water, two sandwiches to share with friends, and comfortable shoes for walking. It turns out $\frac{2}{3}$ of the students forgot to bring a canteen, $\frac{1}{2}$ of the students forgot to bring a sandwich to share, and $\frac{2}{5}$ of the students did not bring appropriate shoes for walking. They all came to the park in two buses that fit 35 students each, and they could not have made it with one single bus. How many students went to the expedition to the park?

Problem 7

There were 3 adults and 6 children lining up for a picture. The photographer decided the picture would look better if there were no two adults standing next to each other in the picture. In how many different ways can they line up to take the picture?

Problem 8

Equilateral triangles and squares are added to the figure below, one at a time, alternating square and triangle.

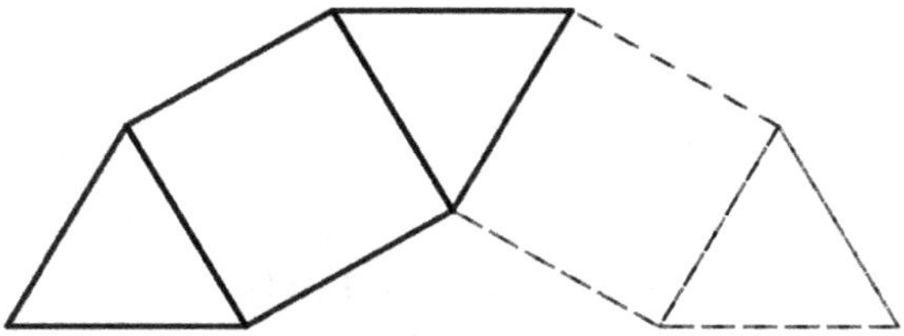

The figure is complete when it is not possible to add another figure without overlapping any of the existing ones. How many squares and triangles would be needed in total to complete the figure?

Problem 9

Chris is training for a 10 km run. While he trains, he runs for 8 minutes, then rests for 2 minutes, runs again for 8 minutes, and rests for 2 minutes. He keeps going until he finishes running 10 km. If he runs at 16 km per hour, how many minutes does he spend training (including the resting times)? Round your answer to the nearest tenth if necessary.

Problem 10

How many factors of 2^{2018} are greater than 2018?

Problem 11

Several equilateral triangles of side length 1 and trapezoids with bases of length 1 and 2 are arranged in a line following the pattern in the figure below.

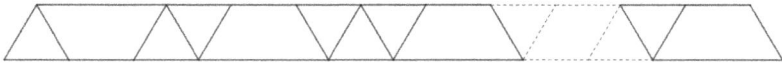

If 35 triangles and trapezoids were used in total, what is the perimeter of the resulting figure?

Problem 12

The mayor of Cincinnati ordered huge signs with the letters of the city (CINCINNATI) to have the posted at the main entrance to welcome visitors. Each of the signs were covered in plastic for protection but this plastic also covered the letters so they all looked identical. The crew in charge of placing the signs did not remove the plastic before posting them. The probability that they placed the letters in the correct order is $\dfrac{P}{Q}$ as a reduced fraction. What is $Q - P$?

Problem 13

Kaila is deciding where to go on vacation. She decided to use a coin to help her decide what to choose. She will throw a fair coin once. If she gets Heads she will go to the Bahamas and if she gets tails she will go to Hawaii. If she goes to the Bahamas she will throw the coin again 3 times in a row and will stay two more nights than the number heads she gets. If she goes to Hawaii she will throw the coin again 2 times and will stay two more nights than twice the number of tails she gets. The probability that she goes to Hawaii for 4 or more nights is $\dfrac{P}{Q}$. What is $Q - P$?

Problem 14

Lady, the dog, is waiting for her master with her leash tied to a sculpture that has a base in the shape of a regular hexagon with side length 2 meters, as shown on the diagram below.

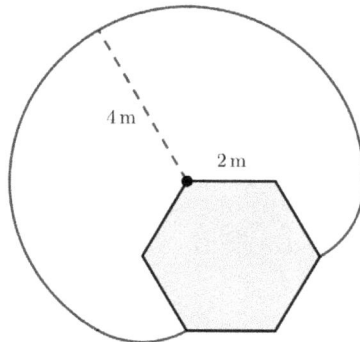

Lady's leash is 4 meters long, so she can freely walk along a region of grass that has area $K \times \pi$ for an integer K. What is K?

Problem 15

Gregor is thinking of two positive integers. He tells his son that the product of the numbers is 8400, and the largest number that divides both is 20. If both numbers are greater than 50, what is the smaller of the numbers Gregor is thinking of?

Problem 16

Andie is on a shopping spree and came across a store that offers great discounts. He will buy 3 identical pairs of jeans with a Buy Two Get One Free offer (that is, he gets 3 for the price of 2), each originally priced at $75, and 2 dress shirts that originally cost $25 with a 20% discount. Altogether Andie saved $P\%$ on his purchase. What is P? Round your answer to the nearest integer.

Problem 17

How many 4-digit integers $\overline{abcd}$ are such that $a + 3b + 3c + d$ is a multiple of 3?

Problem 18

The grid below is made up with squares with side length 5. Four triangles are formed and shaded as in the diagram below.

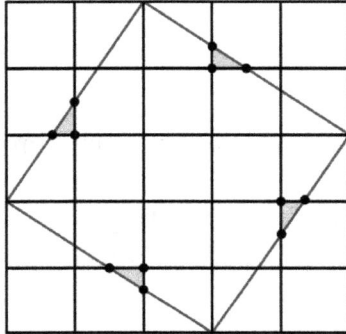

The area of the shaded region is $\dfrac{P}{Q}$ in simplest terms. What is $P+Q$?

Problem 19

A real number between 3 and 7.5 is chosen at random from the number line (note the number is not necessarily an integer). The probability that the square of the number is between 16 and 49 is $\dfrac{P}{Q}$. What is $Q-P$?

Problem 20

A train travels at constant speed for several hours. It takes 32 seconds for it to pass completely through a tunnel which is 356 meters long, and 25 seconds for it to pass completely through a bridge that is 230 meters long. (For example, this means 32 seconds pass from when the train starts to enter the tunnel to when the train is completely out of the tunnel.)

A second train approaches on a parallel track at a speed of 22 meters per second. If it takes 10 seconds for the trains for pass each other completely, how many meters long is the second train?

1.8   ZIML May 2018 Division M

Below are the 20 Problems from the Division M ZIML Competition held in May 2018.
The answer key is available on p.191 in the Appendix.
Full solutions to these questions are available starting on p.153.

Problem 1
In the diagram below *ABCDE* is a regular pentagon, *BAFGHI* is a regular hexagon, and *CIJ* is an equilateral triangle.

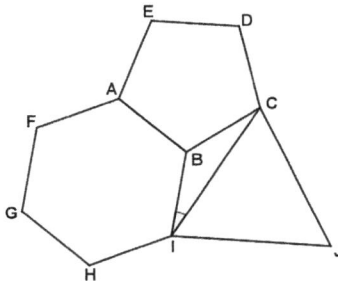

What is the measure of $\angle BIC$ in degrees?

Problem 2
In Ms. Perez's class there are 5 students that did not do the English nor the Spanish homework, 12 students that did not do the English homework, and 12 students that did both the English and Spanish homework. If there are 34 students in Ms. Perez's class, how many students did not do the Spanish homework but did the English homework?

Problem 3

How many numbers of the numbers $20, 21, 22, \ldots, 45$ have exactly 6 factors?

Problem 4

The number of red pens and black pens in Mrs. Darcy's bag are in ratio $3 : 11$. If Mrs. Darcy has 98 pens in her bag, how many of those are red pens?

Problem 5

Judy needs to set up a 5-digit password. She tends to forget things so she wants to make a password that is easy to remember. Judy loves the number 66, so she wants a number that is a multiple of 66, and she will only use the digits 3 and 6 to set up her password. What is Judy's password?

Problem 6

The school bus drops students off at 5 different stops. The bus driver must make a list of how many students left the bus at each stop (he does not need to record which student leaves at a particular spot, just the number). Some days no one leaves at certain stops, so the bus may skip some stops with a 0 recorded in the list. If there are 12 students on the bus, how many different lists could the bus driver have at the end of his trip?

Problem 7

ABCD is a 5×2 rectangle. Isosceles right triangle $\triangle ABE$ is drawn overlapping this rectangle as shown in the diagram below:

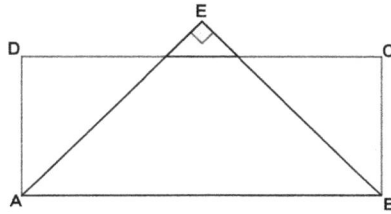

What is the area of the portion of rectangle *ABCD* outside of $\triangle ABE$? Round your answer to the nearest tenth if necessary.

Problem 8

Anastasia can make three dozen tortillas in 20 minutes. Her sister Lucresia can make two dozen tortillas in 10 minutes. They got a big order of 456 tortillas. Anastasia worked alone making tortillas for 90 minutes, then Lucresia joined her and they finished together. How many minutes did Lucresia work making tortillas? Round your answer to the nearest minute if necessary.

Problem 9

Let a *A* be a positive integer with exactly 5 factors, and *B* a positive integer with exactly 6 factors. If $\text{lcm}(A, B) = 324$, what is $A + B$?

Problem 10

For the Cinco de Mayo weekend, La Taqueria will be serving tacos all night on Thursday, Friday, and Saturday, and needs 2 employees to work each night. The 5 employees decide to keep it fair, so each employee will work at least one night shift, with no one working the night shift 2 days in a row. In how many different ways can they decide the schedule of who works Thursday, Friday, and Saturday?

Problem 11

Alex took $360 from the bank. He got $5, $10, and $20 bills, 32 in total. If he got twice as many $10 bills as $20 bills, how many $5 bills did he get?

Problem 12

Robin and Clinton are practicing their archery. They shoot at an 8×8 checkerboard made up of identical squares. Assume each of their shots lands randomly on the checkerboard. Robin shoots first. When Clinton shoots, the probability that his shot lands in the same row or column as Robins can be expressed as $P\%$ for a number P. What is P, rounded to the nearest integer?

Problem 13

In the following diagram the side length of the equilateral triangles is half the side length of the next bigger equilateral triangle.

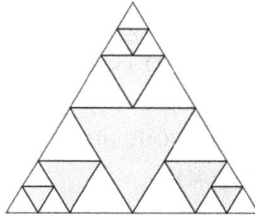

If the shaded area is 93, what is the area of the full (biggest) triangle?

Problem 14

Pretend you live in a society where there are only $5 bills and $9 bills. What is the largest dollar amount that you cannot pay using only these bills (without getting any change back)?

Problem 15

Let $ABCD$ be a parallelogram, and E and F points on its diagonal $\overline{AC}$ such that $AE : EF : FC = 1 : 1 : 2$. If $[ABCD] = 120$, what is $[ABE] + [CFD]$? Round your answer to the nearest integer if necessary. Here, for example, $[ABCD] = 120$ means that the area of parallelogram $ABCD$ is 120.

Problem 16

On my road-trip from San Diego to San Francisco I decided to stop at Santa Monica and Morro Bay. Santa Monica is 134 miles away from San Diego, Morro Bay is 200 miles from Santa Monica, and San Francisco is 232 miles from Morro Bay. If my average speed from San Diego to Santa Monica was 67 mph, from Santa Monica to Morro Bay was 60 mph, and from Morro Bay to San Francisco was 58 mph, what was my average speed for the whole trip? Give your answer in mph and round to the nearest integer if necessary.

Problem 17

What is the remainder of 2018^{2018} after dividing by 5?

Problem 18

Dolores bought a new board game that came with a pair of 6-sided dice. This dice are such that the probability of getting 1 is $\frac{1}{3}$ and the probability of getting each of the other five numbers is $\frac{2}{15}$. The probability that Dolores gets a sum of 5 when she rolls both dice is $\frac{P}{Q}$ as a reduced fraction. What is $Q - P$?

Problem 19

For this problem use the following approximations (you may not need all of them in your calculations):

$$\pi = 3.1, \sqrt{2} = 1.4, \sqrt{3} = 1.7, \sqrt{5} = 2.2.$$

Consider two circles with the same center, one with radius 2 and other with radius 4, as in the diagram below.

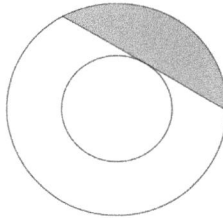

What is the shaded area, rounded to the nearest tenth?

Problem 20

Mr. Lin is a driver for the bus route that goes from city A to city B. Buses on that route leave every 35 minutes from city A towards city B and from city B towards city A, starting at 5:30 AM. Mr. Lin is driving a bus that left the station on city A at 7:15 AM. If it takes a bus 55 minutes to get from city A to city B (or vice versa), how many buses that go from city B to city A is Mr. Lin going to see on his way?

1.9 ZIML June 2018 Division M

Below are the 20 Problems from the Division M ZIML Competition held in June 2018.
The answer key is available on p.192 in the Appendix.
Full solutions to these questions are available starting on p.163.

Problem 1
Curtis bought supplies for his office. He spent $150 in paper. He bought regular letter size paper for $6 per pack, recycled letter size paper for $3.50 per pack, and legal size paper for $6.50 per pack. He bought twice as many packs of regular letter size paper than recycled letter size paper. If he bought 28 packs of paper in total, how many packs of recycled letter size paper did he buy?

Problem 2
Esmeralda works at a jewelry store. She needs to polish and pack some rings for a huge sale. After working for 4 hours the ratio of the number of rings that are ready to the number of rings that she still needs to pack is 4 : 9. After packing 135 more rings the ratio becomes 9 : 4. When Esmeralda finishes, how many rings will she have packed in total?

Problem 3
Leesa is a saleswoman and gets paid by commission. She gets 8% commission on her first $800 of weekly sales and 10% commission on any sales past $800. If Leesa's sales this week were $1530, how many dollars was her commission?

Problem 4

Janis went hiking at a national park. On her way to the top of the hill she traveled at an average speed of 3 miles per hour. On her way back she traveled at an average speed of 4 miles per hour. What was her average speed for the whole trip? Give your answer in miles per hour rounded to the nearest tenth if necessary.

Problem 5

Lilly and Jay work at an ice cream shop. It takes Lilly 6 hours to make 9 liters of ice cream, and it takes Jay 4 hours to make 8 liters of ice cream. How many hours would they need to work together to make 21 liters of ice cream?

Problem 6

In the following diagram the grid is made out of squares of side length 1.

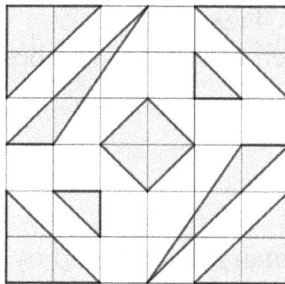

What is the combined area of all figures on the grid?

Problem 7

Consider pentagon $ABCDE$. Let F, G, and H be points outside of the pentagon such that BCF, CDG and EAH are equilateral triangles. What is the measure (in degrees) of the smallest angle in $\triangle FGH$?

Problem 8

Mr. Drizzle is painting a wall using a paint roller. The rolling part of the paint roller is a cylinder with radius 3 inches and it is 15 inches long. Every time he paints, Mr. Drizzle can roll the paint roller until it has turned 5 whole turns, then he needs to soak it in paint again. How many square inches of wall can Mr. Drizzle paint each time he soaks his paint roller in paint? For this problem use the approximation $\pi \approx 3.14$.

Problem 9

In the following diagram the circles have radii 1, 2, and 3.

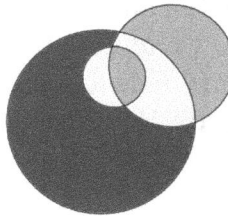

The (positive) difference between the black and gray areas is $K \times \pi$, where K is an integer. What is K?

Problem 10

Donovan drew a rectangle with dimensions $2\sqrt{3}$ and 4. Then he drew two overlapping equilateral triangles inside of the square that shared a side with the shorter side of the rectangle, as shown in the diagram below.

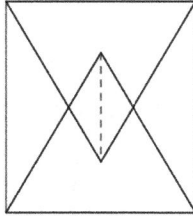

What is the distance between the vertices of the two triangles? Round your answer to the nearest tenth if necessary.

Problem 11

At Mrs. Patel's class there is a rule that says that all students must shake the hand of all other students before the class starts. Yesterday all students were present and 136 handshakes were exchanged. Today, some students were absent and only 91 handshakes were exchanged. How many students missed today's class?

Problem 12

How many numbers less than 10000 are divisible by 3^3 but not by 3^6?

Problem 13

You need to buy 10 sodas for a party. There are 5 types of soda available, with one of the 5 types being Coke. If you randomly choose the flavors of soda (in no particular order), the probability that you buy at least two Cokes is $\dfrac{P}{Q}$ as a reduced fraction. What is $Q - P$?

Problem 14

You are dealt 5 cards from a deck of cards that has 5 black cards, 4 blue cards, and 3 red cards. If all cards are different, in how many ways can you get exactly 3 black cards?

Problem 15

Jessenia was visiting the local fair and came across a game that had a spinning wheel with regions of various points like the one in the diagram below.

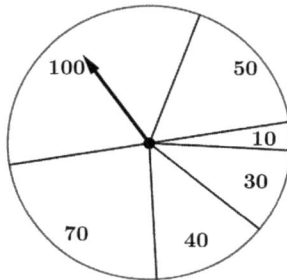

She spins the wheel three times and adds together her points. She wins a major prize if she has 50 or fewer points in total. The size of each of the circular sectors of the wheel is proportional to the number of points it gives, so for example, the region of 30 points is three times larger than the region with 10 points. The probability that Jessenia wins a major prize is $\dfrac{P}{Q}$ as a reduced fraction. What is $Q - P$?

Problem 16

Monica was asked to come up with the smallest number that (i) used only the digits 2, 3 and 5; (ii) was a multiple of 15; and (iii) used at least two 2's, at least two 3's, and at least two 5's as digits. What is Monica's number?

Problem 17

How many 3-digit numbers $\overline{abc} = 100a + 10b + c$ are such that $a + 2b + 3c$ is a multiple of 3?

Problem 18

Sarah wrote down a sequence number according to the following rules. If the current number was odd, she multiplied it by 2 to get the next number. If it was even, she added 3 to it to get the next number. For example, if Sarah started with 4 she would get the sequence $4, 7, 14, 17, 34, \ldots$. If Sarah started with the number 6 and wrote down 100 numbers in total, what was the ones digit of the 100th number?

Problem 19

Jerry always brings a bag with the same number of candy to lunch to share with his friends. Jerry divides the candy evenly among him and all his friends, giving them as many candies as possible and leaving some leftover candy that he eats later at home. Yesterday one of his friends was missing, and he had 4 candies left over. Today all of his friends were there and he had 7 candies left over. If Jimmy shares his candy with at most 10 friends and the bag has less than 40 candies, how many friends does Jerry share his candies with?

Problem 20

A prison has 100 inmates in 100 prison cells, with 100 guards. Each prisoner is in one of the 100 cells and at first all doors are unlocked. In order to lock a door, at least 3 different guards must use their key on that door. The first guard goes through and locks all doors. Then the second guard goes through the prison and locks every second door. The third guard locks every third door. The rest of the guard continue in this manner until all 100 guards have attempted to lock some doors, after which every prisoner whose cell is left unlocked is free to leave. How many prisoners will be set free?

2. ZIML Solutions

This part of the book contains the official solutions to the problems from the nine Division M ZIML Contests from the 2017-18 School Year.

Students are encouraged to discuss and share their own methods to the problems using the Discussion Forum on ziml.areteem.org.

2.1 ZIML October 2017 Division M

Below are the solutions from the Division M ZIML Competition held in October 2017.

The problems from the contest are available on p.17.

Problem 1 Solution

Since 1 ant eats $\frac{1}{15}$ of a cube of sugar in 1 minute, 2 ants will eat $\frac{2}{15}$ of a cube of sugar in 1 minute. So, 2 ants take $\frac{15}{2}$ minutes to eat one whole cube of sugar. Hence, to eat 4 cubes of sugar 2 ants will need $\frac{15}{2} \times 4 = 30$ minutes.

Answer: 30

Problem 2 Solution

The ratio of the amount of work they can complete in a given time is

$$\frac{1}{5} : \frac{1}{3} = 3 : 5.$$

Thus, working for 90 hours, Rachel completes $\frac{3}{8}$ of the project and Anna completes $\frac{5}{8}$ of the project. This means, Anna completes

$$\frac{5}{8} \div 90 = \frac{1}{144}$$

of the work in 1 hour. Hence, Anna needs 144 hours to finish the project alone.

Answer: 144

Problem 3 Solution

Note the big square has the same area as 9 small squares. This means the area of one small square is $32 \div (9 - 1) = 4$, and the area of the big square is $4 \times 9 = 36$. The area of the middle sized squares on the right is $\frac{1}{4}$ of the area of the big square, so each has area $36 \div 4 = 9$. Thus, the area of the whole rectangle is $4 \times 3 + 9 \times 2 + 36 = 66$.

Answer: 66

Problem 4 Solution

Since your average in 7 attempts is 14.9 seconds, the sum of your attempts is $7 \times 14.9 = 104.3$ seconds.

If you want your average to be 14.8 seconds, you want the sum of 8 attempts to be $8 \times 14.8 = 118.4$. Thus, your time on the next attempt needs to be $118.4 - 104.3 = 14.1$ seconds to improve the average to 14.8 seconds.

Answer: 14.1

Problem 5 Solution

The number must be 1 less than a multiple of 5 and 1 less than a multiple of 6, so it should be 1 less than a multiple of 30. The largest multiple of 30 that has three digits is 990, so the number we are looking for is $990 - 1 = 989$.

Answer: 989

Problem 6 Solution

At first Harriet has answered $\frac{4}{9}$ of the questions. After answering 5 more questions, she has answered $\frac{5}{9}$ of the questions. This means those 5 questions are $\frac{5}{9} - \frac{4}{9} = \frac{1}{9}$ of the total questions on the assignment. Thus, Harriet had to answer $5 \times 9 = 45$ questions in total.

Answer: 45

Problem 7 Solution

There are $2^3 = 8$ possible outcomes when flipping a coin 3 times:

$$HHH, HHT, HTH, HTT, THH, THT, TTH \text{ and } TTT.$$

From those, 4 have more tails than heads, so the probability of getting more tails than heads is $\dfrac{4}{8} = \dfrac{1}{2}$. Therefore, $Q - P = 1$.

Answer: 1

Problem 8 Solution

We know there are 12 cars that are neither red nor electric, so there are $40 - 12 = 28$ cars that are red or electric (or both). Since we have 20 red cars and 15 electric cars in the parking lot, $20 + 15 - 28 = 7$ of them are both red and electric.

Answer: 7

Problem 9 Solution

Here we use $[ABCD] = 64$ to denote the area. Since F is the midpoint of DC, $[AFD] = [ABCD]/4 = 16$. Both $\triangle AFD$ and $\triangle EFD$ have the same altitude from F and have bases of equal length, so they have the same area. Therefore the area of the shaded region is $[AFD] + [EFD] = 16 + 16 = 32$.

Answer: 32

Problem 10 Solution

Your club has 7 members and the same person cannot be president and vice-president at the same time. Choosing first the president you have 7 options, and then you have 6 options left to choose the vice-president. Thus, there are $7 \times 6 = 42$ different ways of choosing a president and a vice-president.

Answer: 42

Problem 11 Solution

The cold pump alone fills $\dfrac{1}{17}$ of the pool in one hour, so after the first 5 hours, $\dfrac{5}{17}$ of the pool is full. Together the hot and cold water pumps can fill

$$\frac{1}{17} + \frac{1}{34} = \frac{3}{34}$$

of the pool in one hour. When the maintenance guy opens the second pump, the pool still needs to be filled with $\dfrac{12}{17}$ of its water, so it takes

$$\frac{12}{17} \div \frac{3}{34} = 8$$

hours to finish filling the pool with water. Thus, the cold pump was open for a total of $13 + 5 = 18$ hours.

Answer: 13

Problem 12 Solution

Note $\angle GHC$ and $\angle EHB$ are supplementary, hence $\angle EHB = 180° - 105° = 75°$. Since $\triangle GHC$ is isosceles, $\angle C = 180° - 75° - 75° = 30°$ degrees. Thus, $\angle E = 180° - 105° - 30° = 45°$ degrees.

Answer: 45

Problem 13 Solution

They can choose an appetizer together in 3 ways. Jacky can choose her entree in 10 ways and Jake can also choose his entree in 10 ways. Jacky can choose her dessert in 4 ways, but since Jake wants a different dessert than Jacky, he can choose his dessert in 3 ways. In total they can choose their dinner in $3 \times 10 \times 10 \times 4 \times 3 = 3600$ different ways.

Answer: 3600

Problem 14 Solution

We know the number is divisible by 8, so its last three digits must be divisible by 8. Hence, the digit b should be either 2 or 6, however, the number must not have repeated digits, so $b = 6$. Since the number is divisible by 6, the sum of its digits must be divisible by 3, so a should be either 2, 5 or 8. The only option with no repeated digits is 82560.

Answer: 82560

Problem 15 Solution

The last digit of a power of 2017 is the same as the last digit of a power of 7. The last digits of the first few powers of 7 are 7, 9, 3, 1, 7, 9, ..., which follow a cycle of length 4. Since 2017 leaves remainder 1 when divided by 4, the last digit of 2017^{2017} is the same as the last digit of 7^1, which is 7.

Answer: 7

Problem 16 Solution

Let's first count the triangles ignoring the dotted line in the following diagram:

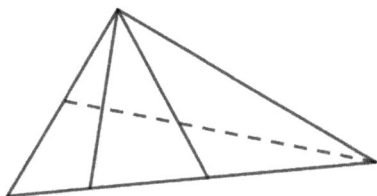

We have 3 small triangles, 2 triangles made of two small triangles, and 1 big triangle, for a total of 6 triangles. Considering the dotted line as well, the same argument as above gives 6 triangles in the top half (with the dotted line as the base). In the bottom half there are 3 additional triangles (made up of 1, 2, and 3 regions). This gives a total of $6 + 6 + 3 = 15$ triangles.

Answer: 15

Problem 17 Solution
Looking at prime factorizations, we have
$$2550 = 2^3 \times 3^2 \times 5 \times 7 \text{ and } 20 = 2^2 \times 3^0 \times 5 \times 7^0.$$
Since $40 = 2^3 \times 3^0 \times 5 \times 7^0$, looking at the factors and exponents of the LCM and GCD, the other number must be $2^2 \times 3^2 \times 5 \times 7 = 1260$.

Answer: 1260

Problem 18 Solution
We can divide the regular hexagon into 6 identical equilateral triangles with the same side length as the hexagon, each of area $54\sqrt{3} \div 6 = 9\sqrt{3}$. Recalling that the area of an equilateral triangle with side length s is $\dfrac{s^2\sqrt{3}}{4}$, we see the side length must be 6.

Answer: 6

Problem 19 Solution
The number of people that visited the store must be divisible by both 3 and 7, so it is a multiple of 21.

The only multiple of 21 between 120 and 144 is 126. Since 1 out of 7 people bought candy corn, $126 \div 7 = 18$ people bought candy corn.

Answer: 18

Problem 20 Solution
Since Rick wants all the physics books to be together, let's pretend he has 1 large physics book, 4 math books and 2 biology books.

Thus, we want to arrange $4 + 1 + 2 = 7$ books. That can be done in $7! = 5040$ ways. Since there are 3 physics books, they can be arranged in $3! = 6$ ways, so there are $5040 \times 6 = 30240$ different ways to arrange all books on the shelf.

Answer: 30240

2.2 ZIML November 2017 Division M

Below are the solutions from the Division M ZIML Competition held in November 2017.

The problems from the contest are available on p.25.

Problem 1 Solution

Notice the shaded area is equal to the area of one of the squares minus two small triangles with base and height $21 - 14 = 7$. Therefore, the shaded area is

$$21 \times 21 - 2 \times \frac{7 \times 7}{2} = 392.$$

Answer: 392

Problem 2 Solution

Rita has 20 red and 16 white marbles, so the ratio of red to white marbles is $20 : 16 = 5 : 4$ for both Rita and Rose.

Therefore, for every 4 white marbles, Rose has 5 red marbles, so $\frac{4}{9}$ of Rose's marbles are white. We thus can calculate that Rose has $27 \times \frac{4}{9} = 12$ white marbles.

Answer: 12

Problem 3 Solution

Jane's parents visit every 6 weeks and grandparents every 8 weeks. They both will visit on weeks that are a common multiple of 6 and 8. Thus they will visit every week that is a multiple of the LCM of 6 and 8, which is 24. Thus both her parents and grandparents will visit Jane during weeks 24 and 48, which is 2 times in 2017.

Answer: 2

Problem 4 Solution

Since the only prime number between 20 and 25 is 23 we need to check which of the numbers $230, 231, 232, \ldots, 239$ are prime. Clearly the even numbers are not, 231 and 237 are divisible by 3, and 235 is divisible by 5. In fact 233 and 239 are prime, so the answer is 2.

Answer: 2

Problem 5 Solution

Since Mark wants to bring at least 3 each day and at least 5 on Friday, he just needs to figure out how to sort the rest of the candy. After saving the minimum amount of candy for each of those days he will have $21 - 3 - 3 - 3 - 3 - 5 = 4$ pieces of candy left. We can use stars and bars to figure out how to arrange those. We can use 4 stars to represent the pieces of candy and $5 - 1 = 4$ bars to represent the days of the week. Therefore, there are

$$\binom{4+5-1}{4} = \binom{8}{4} = 70$$

possible ways to distribute the rest of the candy throughout the week.

Answer: 70

Problem 6 Solution

Since $\frac{1}{4}$ of the students that passed got an A, the number of students that passed must be a multiple of 4. Similarly as $\frac{1}{3}$ of the students that passed got a B, the number of students that passed must also be a multiple of 3.

Hence the number of students that passed is a multiple of $\mathrm{lcm}(3,4) = 12$. As the class has 20 students, the only multiple of 12 that works is 12 itself. Thus, 12 students passed the exam.

Answer: 12

Problem 7 Solution

Jack's brother can clean $\frac{1}{60}$ of the kitchen per minute. When Jack and his brother work together, they can clean $\frac{1}{45}$ of the kitchen per minute. Then, we can find the amount of work that Jack can do per minute

$$\frac{1}{45} - \frac{1}{60} = \frac{1}{180}.$$

Since Jack can clean $\frac{1}{180}$ of the kitchen per minute, it would take him 180 minutes to clean the kitchen alone.

Answer: 180

Problem 8 Solution

First note as $\triangle ABC$ is an isosceles right triangle, $\angle BAC = 45°$. Since $\angle FEB = 165°$, $\angle BED = 180° - 165° = 15°$, and $\angle EBD = 90° - 15° = 75°$. Notice that $\angle CAG = \angle CEG = \angle DEB = 15°$ and $\angle AFG = \angle CFE = \angle DBE = 75°$. Therefore, $\angle GAB = 45° + 15° = 60°$.

Answer: 60

Problem 9 Solution

Extend lines to make a big rectangle with height 8 and length 16. The shaded region is then the area of the big rectangle, minus a 4 by 4 square and two triangles (with respective bases and heights of $16, 4$ and $12, 8$.

The area of the shaded region is therefore

$$16 \times 8 - 4^2 - \frac{1}{2} \times 16 \times 4 - \frac{1}{2} \times 12 \times 8$$
$$= 128 - 16 - 32 - 48$$
$$= 32.$$

Answer: 32

Problem 10 Solution

The first week Jim got 6% of 500 dollars, and 10% of $2000 - 500 = 1500$ dollars, that is, Jim got a bonus of

$$500 \times 6\% + 1500 \times 10\% = 180$$

dollars.

The second week Jim got 6% of 500 dollars, and 10% of $3200 - 500 = 2700$ dollars, that is, Jim got a bonus of

$$500 \times 6\% + 2700 \times 10\% = 300$$

dollars. However, since the second week he sold more than $2500, that week his bonus increased by 20% so he got an extra $300 \times 20\% = 60$ dollars.

In total, Jim got a bonus of $180 + 300 + 60 = 540$ dollars.

Answer: 540

Problem 11 Solution

Note that only perfect squares have an odd number of factors. The perfect squares between 140 and 700 are

$$12^2 = 144, \ 13^2 = 169, \ \ldots, \ 26^2 = 676.$$

Hence there are $26 - 12 + 1 = 15$ numbers between 140 and 700 that have an odd number of factors.

Answer: 15

Problem 12 Solution

$3600 = 2^4 \times 3^2 \times 5^2$, and $7560 = 2^3 \times 3^3 \times 5 \times 7$, thus

$$\gcd(3600, 7560) = 2^3 \times 3^2 \times 5 = 360.$$

This means we are looking for 2-digit factors of 360. These factors are

$$10, 12, 15, 18, 20, 24, 30, 36, 40, 45, 60, 72 \text{ and } 90.$$

Thus, there are 13 factors.

Answer: 13

Problem 13 Solution

The passenger gets on the bus at 6:24 AM to travel to station B. Since the one-way trip takes 52 minutes, none of the buses from station B have arrived at station A when the passenger departs. The passenger will therefore see every bus that departs station B between 6 AM and their arrival time at station B which is 7:16 AM. One bus leaves exactly at 6 AM, and in the 76 minutes that follow 6 more buses leave, because

$$76 \div 12 \approx 6.3$$

and we round down because the buses leave at the end of each 12 minute interval. Therefore, the passenger sees a total of

$$6 + 1 = 7$$

buses en route from station A to station B.

Answer: 7

Problem 14 Solution
Notice

$$[ADC] = [BDC] = \frac{24 \times 8}{2} = 96$$

as both $\triangle ADC$ and $\triangle BDC$ have the same base and height. Each of the shaded regions can be obtained by removing $\triangle EDC$ from $\triangle AED$ and $\triangle BDC$, respectively, thus

$$[AED] = [BEC] = [ADC] - [EDC] = 32.$$

Therefore, the shaded region has area $2 \times 32 = 64$.

Answer: 64

Problem 15 Solution
Assuming the letters are all distinct, there are

$$9! = 9 \times 8 \times 7 \times 6 \times 5 \times 4 \times 3 \times 2 \times 1 = 365880$$

ways to arrange 9 letters to form words. However, since there are two Ls and two Es, we need to remove all duplicate arrangements by dividing by the number of ways to arrange the two Ls and also by the ways to arrange the two Es, each of which can be done in 2 ways. Therefore, there are

$$\frac{365880}{2 \times 2} = 90720$$

ways to arrange the letters in the word "HALLOWEEN".

Answer: 90720

Problem 16 Solution

Since $\overline{5b}$ is divisible by 3, b should be either 1, 4 or 7. If $\overline{32a5b}$ is to be divisible by 11, it must be the case that

$$3 - 2 + a - 5 + b = a + b - 4$$

is a multiple of 11. If $b = 1$, then $a = 3$; if $b = 4$ then $a = 0$; and if $b = 7$, then $a = 8$. The three 4-digit numbers that satisfy these properties are then 32351, 32054 and 32857. Thus, the largest 5-digit number with these properties is 32857.

Answer: 32857

Problem 17 Solution

Note $360° \div 45° = 8$, so the sector is one eighth of a circle with radius 42. The sector there has area

$$\frac{1}{8}\pi \times 42^2 \approx \frac{1}{8} \times \frac{22}{7} \times 42^2 = 693.$$

The triangle is an isosceles right triangle with diagonal 42, so it has area

$$\frac{1}{4} \times 42^2 = 441.$$

Hence the shaded region has area

$$L = 693 - 441 = 252.$$

Answer: 252

Problem 18 Solution

If we subtract 7 from the number we are looking for, it must be a multiple of all 8, 10 and 12. Hence, the number we are looking for is 7 more than a multiple of $\text{lcm}(8, 10, 12) = 120$.

The largest 3-digit number that is a multiple of 120 is $120 \times 8 = 960$. Therefore, the number we are looking for is $960 + 7 = 967$.

Answer: 967

Problem 19 Solution

The whole number line has length $14 - 2 = 12$. You win at least as much as you paid if the dial is between 5 and 14 which has length $14 - 5 = 9$. Thus the probability is

$$\frac{9}{12} = \frac{3}{4} = 75\%,$$

so our answer is 75.

Answer: 75

Problem 20 Solution

We can compare the differences of the original concentration from each solution with the target concentration. The differences are $15 - 10 = 5$ and $30 - 15 = 15$, respectively. The ratio of this differences is $5 : 15 = 1 : 3$. In order to make the concentration of the mixture 15%, we need the number of gallons of each solution to have the opposite ratio.

That is, the ratio of the number of gallons of the first solution to the number of gallons of the second solution must be $3 : 1$, which is the same as $150 : 50$. So, if we mix 150 gallons of the first solution and the 50 gallons of the second solution, we will get a solution that is 15% ammonia.

Answer: 150

2.3 ZIML December 2017 Division M

Below are the solutions from the Division M ZIML Competition held in December 2017.
The problems from the contest are available on p.33.

Problem 1 Solution
Since Grace and Danny are traveling in opposite directions, their relative speed is the sum of the two speeds, which is, $5 + 7 = 12$ km per hour towards each other. We can then find the time, which is $60 \div 12 = 5$ hours until they pass each other.

Answer: 5

Problem 2 Solution
For the average of the numbers to be 5, we need the sum of the numbers to be $5 \times 7 = 35$. As the sum of the current numbers is 29, if the next number is $35 - 29 = 6$ the average of all numbers will be 5.

Answer: 6

Problem 3 Solution
The left and right sides of the long parallelogram each has length 3 cm, so the top and bottom sides each has length

$$(438 - 2 \times 3) \div 2 = 216$$

cm. Each group of 2 rhombi and 2 triangles has top side length 9 cm, so as $216 \div 9 = 24$ there are 24 full groups. Hence there are $2 \times 24 = 48$ total rhombi.

Answer: 48

Problem 4 Solution
If all 24 motorcycles are the first kind with a capacity of 2 people,

the motorcycles will fit

$$24 \times 2 = 48$$

people, meaning that

$$68 - 48 = 20$$

people will left out. The second kind of motorcycle can hold

$$3 - 2 = 1$$

extra person, so if we switch 20 of the motorcycles to the second kind everyone will fit. Hence we have

$$24 - 20 = 4$$

motorcycles of the first kind and 20 of the second. Since the first kind costs \$40 per motorcycle and the second kind costs \$30, the total cost is

$$4 \times 40 + 20 \times 30 = 760$$

dollars to rent the 24 motorcycles.

Answer: 760

Problem 5 Solution
If we were to list all of the eleven numbers in order, the number right in the middle of the list (the sixth number) would be the average of all of the numbers. Since the sum of the numbers is 616, their average is $616 \div 11 = 56$. Counting backwards by 2 we can see that the second number is $56 - 2 - 2 - 2 - 2 = 48$.

Answer: 48

Problem 6 Solution

The test in total has

$$12 + 15 + 18 = 45$$

questions. For Jong-Zhi to get 75% of them correct, she needs to get

$$75\% \times 45 = 0.75 \times 45 = 33.75$$

questions correct. Thus, to pass Jong-Zhi must get at least 34 questions correct. We know she got 75% of the arithmetic questions correct. Since there are 12 questions, she got

$$75\% \times 12 = 0.75 \times 12 = 9$$

arithmetic questions correct. Similarly, she got 60% of the 15 algebra questions correct which is an additional

$$60\% \times 15 = 9$$

correct questions. Therefore Jong-Zhi has answered

$$9 + 9 = 18$$

questions correct so far. Therefore she needs to answer

$$34 - 18 = 16$$

of the 18 geometry questions to get a passing grade on the test.

Answer: 16

Problem 7 Solution

We note that the bag contains objects defined by its shape (cubes vs. balls) and color (red vs. blue).

Shape and color are independent from each other, so one can interpret the complement of cubed-shaped objects to be ball-shaped objects and the complement of red objects to be blue.

Since we are interested in minimizing the number of red balls in the bag, we want to maximize the number of red cubes in the bag.

Given that there are at least one blue cube in the bag, the maximum number of red cubes is

$$20 - 1 = 19.$$

Since there are 40 total red objects, the minimum number of red balls is

$$40 - 19 = 21.$$

Answer: 21

Problem 8 Solution
Notice that the sequence is

$$0^3 + 2, 1^3 + 2, 2^3 + 2, 3^3 + 2, 4^3 + 2, 5^3 + 2,$$

so the next term is

$$6^3 + 2 = 218.$$

Answer: 218

Problem 9 Solution
Call the entire triangle ADE (with C on $\overline{AE}$). Then $\triangle ADE$ has area 25, denoted $[ADE] = 25$. We know $\triangle ABC$ and $\triangle ABE$ have the same height, so

$$[ABC] : [ABE] = 3 : 5 = 12 : 20.$$

Similarly,

$$[ABE] : [ADE] = 4 : 5 = 20 : 25.$$

Combining these two we have

$$[ABC] : [ADE] = 12 : 25.$$

As $[ADE] = 25$, we have $[ABC] = 12$.

Answer: 12

Problem 10 Solution

Let A, B denote the last two digits so our numbers are of the form $\overline{2017AB}$. Since the number is divisible by 5 we know $B = 0$ or $B = 5$. For a number to be divisible by 3 the sum of its digits must also be divisible by 3. If $B = 0$ the sum of the digits is

$$2 + 0 + 1 + 7 + A + 0 = A + 10 \text{ so } A = 2, 5, 8.$$

Similarly if $B = 5$ the sum of the digits is

$$2 + 0 + 1 + 7 + A + 5 = A + 15 \text{ so } A = 0, 3, 6, 9.$$

In total this gives $3 + 4 = 7$ possible numbers.

Answer: 7

Problem 11 Solution

Let the 20 balls represent the stars/stones and let the $4 - 1$ bars/sticks divide the stars into 4 groups, with each group indicating the number of balls in each box.

If we require at least one ball in each bin, we can reserve 4 stars so that each star will go in each bin. There will be 16 remaining stars.

Therefore, the number of ways to arrange the stars and bars is equal to

$$\binom{16 + 4 - 1}{16} = \frac{19 \times 18 \times 17}{3 \times 2 \times 1} = 19 \times 17 \times 3 = 969.$$

Answer: 969

Problem 12 Solution

The big circle has radius 3. The perimeter of the shaded region equals the sum of the perimeters of all the circles. Therefore the answer is $7 \times (2\pi) + (2 \times 3)\pi = 20\pi$. Thus, $K = 20$.

Answer: 20

Problem 13 Solution

$702 = 2 \times 3^3 \times 13$, so for $702 \times K$ to be a perfect square we need $K = 2 \times 3 \times 13 = 78$.

Answer: 78

Problem 14 Solution

First use the Pythagorean theorem to find the distance between opposite corners of the base (i.e. the length of the diagonal of the rectangular base):

$$\sqrt{3^2 + 4^2} = 5.$$

Note then that this diagonal is perpendicular to the vertical edges of the rectangular prism, so we can use the Pythagorean theorem again to find the distance between opposite corners:

$$\sqrt{5^2 + 12^2} = 13.$$

Therefore the length of the line connecting the two corners is 13.

Answer: 13

Problem 15 Solution

We don't know the length of the trail, or the time she spent uphill or downhill, so we may assume that she took 3 hours going uphill.

Then the trail length was $3 \times 4 = 12$ miles, and the downhill journey took $12 \div 6 = 2$ hours. Therefore the total time was $3 + 2 = 5$ hours, and the average speed of the whole trip was

$(12+12) \div 5 = 4.8$ miles per hour.

Answer: 4.8

Problem 16 Solution

In total there are $\binom{10}{5} = 252$ ways to get 5 cards. There are $\binom{5}{3} \times \binom{5}{2} = 100$ ways to get 3 red and 2 black cards. Hence the probability is

$$\frac{100}{252} = \frac{25}{63}$$

so $Q - P = 63 - 25 = 38$.

Answer: 38

Problem 17 Solution

We can split the triangle into two right triangles:

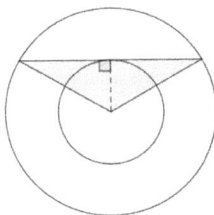

Each right triangle will have one leg of length 1 and hypotenuse of length 2. Notice each of this is half an equilateral triangle, so the shaded area is the same as the area of a an equilateral triangle with side length $\sqrt{2}$. Thus, the shaded area is

$$\frac{\sqrt{3} \times 2^2}{4} = \sqrt{3} \approx 1.7.$$

Answer: 1.7

Problem 18 Solution

The ratio of Harry Potter fans to Twilight fans is equal to $5 : 4 = 3000 : 2400$, so there were 3000 Harry Potter fans at the comic-con. The ratio of Star Trek fans to Harry Potter fans is equal to $5 : 3 = 5000 : 3000$, so there were 5000 Star Trek fans at the comic-con.

Answer: 5000

Problem 19 Solution

Let's look for a pattern in the remainders of powers of 7 and powers of 5 when dividing by 6.

		Remainder
7^1	7	1
7^2	49	1
7^3	343	1
7^4	2401	1

		Remainder
5^1	5	5
5^2	25	1
5^3	125	5
5^4	625	1

Note that every power of 7 leaves remainder 1 when dividing by 6, and every odd power of 5 leaves remainder 5 when dividing by 6. So, the sum $7^{2017} + 5^{2017}$ leaves remainder 0 when dividing by 6.

Answer: 0

Problem 20 Solution

For the first shift on each night all of the $6 + 1 = 7$ people are eligible for the first shift, so there are $\binom{7}{2} = 21$ ways of choosing the first shift. For the second shift, the first two people are not eligible, so there are now $\binom{5}{2} = 10$ ways of choosing who keeps watch. As the second night has the same choices as the first night, there are

$$(21 \times 10)^2 = 210^2 = 44100$$

total ways to decide the 4 shifts.

Answer: 44100

2.4   ZIML January 2018 Division M

Below are the solutions from the Division M ZIML Competition
held in January 2018.
The problems from the contest are available on p.41.

Problem 1 Solution
On the first test Dakota got

$$20 \times 85\% = 20 \times 0.85 = 17$$

correct questions. On the second test he got

$$30 \times 70\% = 30 \times 0.7 = 21$$

correct questions. Thus, he got $17 + 21 = 38$ correct questions
out of $20 + 30 = 50$ questions, that is, he got

$$38 \div 50 = 0.76 = 76\%$$

of the questions correct.

Answer: 76

Problem 2 Solution
Since AB is parallel to CD, we have $\triangle ABE \sim \triangle CDE$. The ratio
of the areas of the triangles is the square of the ratio of their sides.
The ratio of the sides of the triangles is $10 : 15 = 2 : 3$, so their
areas are in ratio $2^2 : 3^2 = 4 : 9 = 36 : 81$. This means the area of
$\triangle CDE$ is 81.

Answer: 81

Problem 3 Solution
Any number that divides both 160 and 60, should also divide their
greatest common factor.

The prime factorizations of our numbers are $160 = 2^5 \times 5$, and $60 = 2^2 \times 3 \times 5$, so their greatest common factor is $2^2 \times 5 = 20$. The divisors of 20 are 1, 2, 4, 5, 10, and 20, so there are 6 positive integers that divide both 160 and 60.

Answer: 6

Problem 4 Solution

Since Stephanie and Bob are traveling in opposite directions, their relative speed is the sum of the two speeds, which is,

$$4 + 6 = 10 \text{ km per hour}$$

towards each other. We can then find the time, which is

$$34 \div 10 = 3.4 \text{ hours}$$

until they pass each other.

Answer: 3.4

Problem 5 Solution

We have $72 \div 7 = 10$ with remainder 2 and $53 \div 7 = 7$ with remainder 4. Hence there are $10 - 7 = 3$ multiples of 7 from 54 to 74.

As there are $72 - 54 + 1 = 19$ numbers between 54 and 72, the probability of choosing a multiple of 7 is $\frac{3}{19}$. Therefore, $Q - P = 19 - 3 = 16$.

Answer: 16

Problem 6 Solution

The hexagon can be divided into 6 congruent triangles.

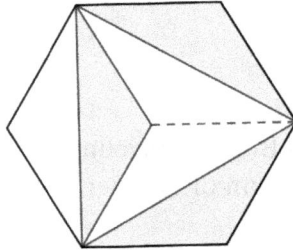

3 of those triangles make up the shaded area. Thus, the shaded area is half of the area of the hexagon, that is, $114 \div 2 = 57$.

Answer: 57

Problem 7 Solution

The prime factorization of 720 is $2^4 \times 3^2 \times 5$. There is only one way to group these factors in 3 groups to obtain consecutive numbers: $8 \times 9 \times 10$. Thus, the pupils ages are 8, 9, and 10 so the sum of their ages is $8 + 9 + 10 = 27$.

Answer: 27

Problem 8 Solution

There are $4 + 5 = 9$ cards in total, so there are $\binom{9}{2} = 36$ ways of choosing 2 of them. There are $\binom{4}{2} = 6$ ways of choosing two blue cards, and $\binom{5}{2} = 10$ ways of choosing 2 red cards. Thus, there are $36 - 6 - 10 = 20$ ways of choosing cards of different colors. Therefore, the probability of getting cards of different colors is $\dfrac{20}{36} = \dfrac{5}{9}$. Hence, $Q - P = 9 - 5 = 4$.

Answer: 4

Problem 9 Solution

The shaded area is equal to the sum of the areas of the three squares minus the area of the two unshaded triangles. The two unshaded triangles are similar, and the ratio of their sides is $3 : 6 = 1 : 2$, thus their heights are 1 and 2. The sum of the area of the triangles is

$$\frac{3 \times 1}{2} + \frac{6 \times 2}{2} = 7.5.$$

Therefore, shaded area is

$$1^2 + 2^2 + 3^2 - 7.5 = 1 + 4 + 9 - 7.5 = 6.5.$$

Answer: 6.5

Problem 10 Solution

Any number that is a multiple of 3, 5 and 7 should also be a multiple of their least common multiple, $3 \times 5 \times 7 = 105$.

The largest 4-digit number that is a multiple of 105 is $95 \times 105 = 9975$, which has repeated digits. We can see $94 \times 105 = 9870$ has no repeated digits, so this is the number we are looking for.

Answer: 9870

Problem 11 Solution

The interior angles of a triangle add up to $180°$, and vertical angles are equal, so

$$\angle GEF = 180° - 85° - 15° = 80°$$

and
$$\angle DCE = 180° - 24° - 80° = 76°.$$

Thus, $\angle ABC = 180° - 76° - 47° = 57°$.

Answer: 57

Problem 12 Solution

Peter paints $\frac{1}{3}$ of the room and Richard paints $\frac{2}{5}$ of the room, so

$$\frac{1}{3} + \frac{2}{5} = \frac{11}{15}$$

of the room is painted and $\frac{4}{15}$ left to paint. Since it takes Peter 1 hour and 40 minutes, or $1\frac{2}{3} = \frac{5}{3}$ hours to finish painting the room, Peter can paint

$$\frac{4}{15} \div \frac{5}{3} = \frac{4}{25}$$

of the room in one hour. Therefore it takes Peter $\frac{25}{4} = 6.25$ hours to paint the entire room.

Answer: 6.25

Problem 13 Solution

The number we are looking for is 1 less than a multiple of 5 and 1 less than a multiple of 6. All multiples of 5 and 6 are multiples of 30, so we are looking for the largest 2-digit number that is 1 less than a multiple of 30. The largest 2-digit number that is a multiple of 30 is 90, so the number we are looking for is $90 - 1 = 89$.

Answer: 89

Problem 14 Solution

The smallest 6-digit number with no repeated digits that starts with 9 is 901234. By looking at the alternating sum of its digits, $9 - 0 + 1 - 2 + 3 - 4 = 7$ we see it is not a multiple of 11. To justify finding the smallest number, let the number be $\overline{9012ab}$. To be a multiple of 11,

$$9 - 0 + 1 - 2 + a - b = 8 + a - b$$

must be a multiple of 11. If $8 + a - b = 0$ we would have $a = 0, b = 8$ or $a = 1, b = 9$ which gives us repeated digits. As it is impossible for $8 + a - b = 22$ we must have $8 + a - b = 11$. Therefore $a - b = 3$ and the smallest choice for a, b with no repeated digits is $a = 6, b = 3$. This gives the number 901263.

Answer: 901263

Problem 15 Solution

If we first assume all 21 bugs are fireflies, then there will be

$$21 \times 6 = 126$$

legs. There are in fact 138 legs, which is

$$138 - 126 = 12$$

more than if we all bugs were fireflies. Each cicada has the same number of legs but each spider has

$$8 - 6 = 2$$

extra legs, so there must be

$$12 \div 2 = 6$$

spiders in total. The number of fireflies and cicadas is thus

$$21 - 6 = 15.$$

Again, assume these 15 bugs are all fireflies. There will be

$$15 \times 2 = 30$$

pairs of wings. In fact, we only have 23 pairs, a difference of

$$30 - 23 = 7$$

pairs of wings. Each firefly has

$$2 - 1 = 1$$

more pair of wings than a cicada, so there must be

$$7 \div 1 = 7$$

cicadas. Hence the number of fireflies is

$$15 - 7 = 8.$$

Answer: 8

Problem 16 Solution

Since there can be at most 3 pigeons in each pigeonhole, there will be either (i) 3 pigeonholes with 3 pigeons and 1 pigeon hole with 1 pigeon, or (ii) 2 pigeonholes with 3 pigeons and 2 pigeonholes with 2 pigeons. For the first case, we need to pick which of the 4 pigeonholes will have only 1 pigeon. This can be done in 4 ways. For the second case, we need to pick which 2 of the 4 pigeonholes will have 2 pigeons. This can be done in $\binom{4}{2} = 6$ ways. Thus, there are $4 + 6 = 10$ ways to decide how many pigeons to put in each of the 4 pigeonholes.

Answer: 10

Problem 17 Solution

After rotating $60°$, $120°$, $180°$, $240°$, and $300°$ from the center, we can see the triangles that should be shaded are

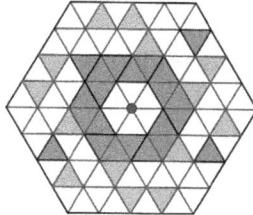

Counting, there are an additional 14 triangles that should be shaded.

Answer: 14

Problem 18 Solution

There are 4 ways of going from city A back to city A: $A \to B \to C \to A$, $A \to C \to B \to A$, $A \to B \to A$ and $A \to C \to A$. For the first case, there are 3 choices to go from A to B, 4 choices to go from B to C, and 3 choices to go from C to A. Thus, there are

$$3 \times 4 \times 3 = 36$$

ways of taking the loop $A \to B \to C \to A$. Similarly, there are

$$2 \times 4 \times 1 = 8$$

ways of taking the loop $A \to C \to B \to A$,

$$3 \times 1 = 3$$

ways of taking the loop $A \to B \to A$, and

$$2 \times 3 = 6$$

ways of taking the loop $A \to C \to A$. Therefore, there are $36 + 8 + 3 + 6 = 53$ loops from city A back to city A.

Answer: 53

Problem 19 Solution

We know that the original ratio of apples to bananas is $3 : 1$ and that 600 pounds of apples are sold every day. Since

$$3 : 1 = 3 \times 200 : 1 \times 200 = 600 : 200$$

if we assume that only 200 pounds of bananas were sold each day, then the amounts of apples and bananas in the warehouse would always be kept at the ratio $3 : 1$. Then at the end, there were 750 pounds of apples, so there would have been

$$750 \div 3 = 250$$

pounds of bananas remaining. Since each day

$$250 - 200 = 50$$

pounds less of bananas were sold than the actual amount, the number of days was

$$250 \div 50 = 5.$$

Thus at the beginning there were

$$250 \times 5 = 1250$$

pounds of bananas, and

$$1250 \times 3 = 3750$$

pounds of apples.

Answer: 3750

Problem 20 Solution

There are 2 ways of adding up to 5 using 3 numbers:

$$1 + 1 + 3 \text{ and } 1 + 2 + 2.$$

Since the dice are of different colors, it matters which number we get on each of them.

There are 3 ways of getting the sum $1+1+3$ (pick one of the three to be the one with the number 3), and 3 ways of getting the sum $1+2+2$ (pick one of the three to be the one with the number 1). Thus, there are 6 different outcomes that add up to 5.

Answer: 6

2.5   ZIML February 2018 Division M

Below are the solutions from the Division M ZIML Competition held in February 2018.
The problems from the contest are available on p.49.

Problem 1 Solution
A number has an odd number of squares if and only if it is a perfect square. So, we are looking for the largest 2-digit perfect square, that is $9^2 = 81$.

Answer: 81

Problem 2 Solution
Note $160 is the final price after a 20% discount, that is, $100\% - 20\% = 80\%$ of the original price. Thus, without a discount Dev would have paid

$$160 \div 80\% = 160 \div 0.8 = 200$$

dollars. With the deal offered by the store, he would have have paid

$$200 \times 90\% \times 90\% = 200 \times 0.9 \times 0.9 = 162$$

dollars. This means Dev paid $162 - 160 = 2$ less dollars than he should have.

Answer: 2

Problem 3 Solution
The largest 5-digit number that can be formed using these 5 digits is 54321, however, $5 + 4 = 9$ is not prime.

$5 + 3 = 8$ isn't prime either, so if we want the first digit to be 5, the second digit must be 2.

$2 + 4 = 6$ is not prime, but $2 + 3 = 5$ is. Note $3 + 4 = 7$ and $4 + 1 = 5$ are all prime numbers, and thus the number 52341 is the largest 5-digit number we can make using the given digits such that the sum of every two consecutive digits is prime.

Answer: 52341

Problem 4 Solution

Since the area of the square is 16, its side length must be 4. Then the rectangle is $4 \times 2 = 8$ units long and $16 \div 8 = 2$ units wide. Thus, the perimeter of the square is $4 \times 4 = 16$ and the perimeter of the rectangle is $2 \times (2+8) = 20$. Therefore, the ratio of the perimeter of the square to the perimeter of the rectangle is $16 : 20 = 4 : 5$, and $a + b = 9$.

Answer: 9

Problem 5 Solution

Since they are running at the same speed, and Larry runs for twice as much time as Moe, Larry ran twice as much distance as Moe. This means Larry ran for 1 mile and Mow ran for .5 miles. Thus, Larry ran for $1 \div 10$ hours, that is, 6 minutes.

Answer: 6

Problem 6 Solution

There are $\binom{15}{5} = 3003$ possible ways of choosing 5 cards at random. There are

$$5 \times \binom{3}{3} \times 4 \times \binom{3}{2} = 5 \times 4 \times 3 = 60$$

different ways of choosing 3 cards with the same number and then 2 other cards with the same number. Thus, the probability is $\dfrac{60}{3003} = \dfrac{20}{1001}$. Hence $Q - P = 1001 - 20 = 981$.

Answer: 981

Problem 7 Solution

All 7 rectangles have the same area, so one rectangle has area $84 \div 3 = 28$. This means the square has area $28 \times 7 = 196$, so it has side length $\sqrt{196} = 14$. Thus, the small rectangles are 14 units long and $14 \div 7 = 2$ units wide.

Answer: 2

Problem 8 Solution

The numbers between 40 and 90 that leave a remainder of 3 when divided by 7 are

$$45, 52, 59, 66, 73, 80, \text{ and } 87,$$

and the ones that leave a remainder of 7 when divided by 13 are

$$46, 59, 72, \text{ and } 85.$$

We can see the only number in both lists is 59, so that is the number we are looking for.

Answer: 59

Problem 9 Solution

Since the distance from A to B is not given, we can assume it is a convenient number. Thus, assume the distance from A to B is 120 km. Therefore, the round trip is 240 km, so if Tom averages a speed of 48 km per hour the round trip will take

$$240 \div 48 = 5$$

hours. Similarly, the drive from A to B takes

$$120 \div 40 = 3$$

hours. Therefore the return trip must take Tom

$$5 - 3 = 2$$

hours. To travel 120 km in 2 hours, Tom must travel

$$120 \div 2 = 60$$

km per hour on the return trip.

Answer: 60

Problem 10 Solution

If you decide to go with 1 scoop of ice cream, you have 3 choices of flavor; if you decide to go with 2 flavors, there are again 3 more choices (you have 3 choices of which flavor NOT to get).

Thus there are $3 + 3 = 6$ ways of choosing the ice cream flavor. As there are 2 choices for the type of cone, and 2 choices for sprinkles (yes or no), there are $2 \times 6 \times 2 = 24$ different ways to customize your ice cream.

Answer: 24

Problem 11 Solution

Since AC is a diagonal of the parallelogram, $\triangle AED$ and $\triangle AEB$ have the same height, and thus, the same area. The shaded area is then the same as the area of $\triangle ADC$, which is half of the area of the parallelogram. Therefore, the area of $ABCD$ is $24 \times 2 = 48$.

Answer: 48

Problem 12 Solution

Considering all tiles to be different, there are 7! different ways of taking out the tiles from the bag.

Of those, there are $2! \times 2!$ arrangements that will give Liam the word CABBAGE, as the 2 As and the 2 Bs can be swapped around. Thus, the probability that he forms the word CABBAGE is

$$\frac{2! \times 2!}{7!} = \frac{2 \times 2}{7 \times 6 \times 5 \times 4 \times 3 \times 2} = \frac{1}{1260},$$

and $P + Q = 1 + 1260 = 1261$.

Answer: 1261

Problem 13 Solution

If the phone is on but not in use, it uses $\frac{1}{18}$ of the battery per hour.
If Carlos is playing, it uses $\frac{1}{4}$ of the battery per hour. Since the
phone has been on for 5 hours, during which he was playing for
2 hours (so not playing for 3 hours), Carlos has used

$$3 \times \frac{1}{18} + 2 \times \frac{1}{4} = \frac{2}{3}$$

of the battery. He therefore has $\frac{1}{3}$ of the battery left, which will
last

$$\frac{1}{3} \div \frac{1}{18} = 6$$

more hours if the phone is on but not playing Pokemon Go.

Answer: 6

Problem 14 Solution

This is equivalent to counting the number of ways to place 3
identical balls in 6 different boxes. Using stars and bars, there

$$\binom{3+6-1}{3} = 56$$

different possible outcomes.

Answer: 56

Problem 15 Solution

Note $\triangle EBF$ is isosceles, and

$$\angle EBF = 360° - 90° - 60° - 60° = 150°.$$

Therefore $\angle BEF = (180° - 150°) \div 2 = 15°$.

Answer: 15

Problem 16 Solution

We can factor $24 = 2^3 \times 3$, so 24 has $(3+1) \times (1+1) = 4 \times 2 = 8$ factors.

Note all factors come in pairs, $1 \times 24 = 24$, $2 \times 12 = 24$, $3 \times 8 = 24$, and $4 \times 6 = 24$. Thus, the product of all factors of 24 is $24^4 = 331776$.

Answer: 331776

Problem 17 Solution

We know the total sale was \$2570. Since \$1970 of this was for cherry and watermelon, the remaining

$$2570 - 1970 = 600$$

dollars were due to strawberry. Hence

$$600 \div 25 = 24$$

kilograms of strawberry candy were sold. Thus the other

$$100 - 24 = 76$$

kilograms were cherry and watermelon. If these 76 kg were all cherry, the total sales would be

$$76 \times 20 = 1520$$

dollars, which is

$$1970 - 1520 = 450$$

less than the actual amount. As each kg of watermelon is

$$30 - 20 = 10$$

dollars per kg more expensive, there must have been

$$450 \div 10 = 45$$

kg of watermelon flavor candies sold.

Answer: 45

Problem 18 Solution
We consider cases based on the shortest stick.

If the shortest stick Roy uses is 2 inches, he can make triangles with sides $(2,3,4)$, $(2,4,5)$, and $(2,5,6)$.

If the shortest is 3 inches, he can make triangles with sides $(3,4,5)$, $(3,4,6)$, and $(3,5,6)$.

Lastly if the shortest is 4 inches, he can make a $(4,5,6)$ triangle.

In total this is $3 + 3 + 1 = 7$ triangles.

Answer: 7

Problem 19 Solution
This is the same as finding the units digit of $5^{33} + 53^{33}$. 5 raised to any power has units digit 5. The powers of 3 form the pattern of

$$3, 9, 7, 1, 3, 9, 7, 1, \ldots$$

for their units digit. As $33 = 32 + 1$, the units digit of 53^{33} is 3. Therefore $35^{33} + 53^{33}$ has units digit $5 + 3 = 8$.

Answer: 8

Problem 20 Solution
We can pretend your parents are one single person, and also that the couple that wants to sit together are one single person. There

are $6! \div 6 = 5! = 120$ ways of sitting 6 people around a circular table.

Now, there are 2! ways to decide how to sit each of the couples, so there are in total $120 \times 2 \times 2 = 480$ different ways that you can decide how to sit around the table.

Answer: 480

2.6   ZIML March 2018 Division M

Below are the solutions from the Division M ZIML Competition held in March 2018.

The problems from the contest are available on p.55.

Problem 1 Solution
The list of numbers that the students have so far, in increasing order, is

$$1, 2, 3, 4, 5, 5, 6, 7, 7, 8, 9, 10, 10,$$

so its median is 6. If Percival picks a number between 1 and 5 the median would be now 5.5, if he picks 6, the median would stay the same, and if he picks a number between 7 and 10, the median would increase to 6.5.

Thus, the probability that the number he picks does not increase the median is $\dfrac{6}{10} = \dfrac{3}{5}$. therefore, $Q - P = 5 - 3 = 2$.

Answer: 2

Problem 2 Solution
Since the area of the circle is π, it must have radius 1. We can split the bigger square into 16 congruent right triangles as shown in the diagram:

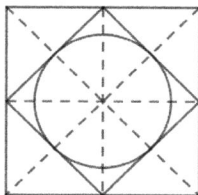

Each of these triangles has area $\dfrac{1}{2}$, so the area of the whole square

---

is $16 \times \dfrac{1}{2} = 8$.

Answer: 8

Problem 3 Solution

We can factor $30 = 2 \times 3 \times 5$, so the fraction will be irreducible if the numerator is not divisible by 2, 3 and 5. There are 8 such numerators:

$$\frac{1}{30}, \frac{7}{30}, \frac{11}{30}, \frac{13}{30}, \frac{17}{30}, \frac{19}{30}, \frac{23}{30}, \frac{29}{30},$$

so 8 of the fractions are irreducible.

Answer: 8

Problem 4 Solution

Since there are 200 Math students and 15% of them are in their 4th year, there are

$$200 \times 15\% = 200 \times 0.15 = 30$$

students in their 4th year. Since there are 500 Computer Science students, and 22% of them are in their 4th year, there are

$$500 \times 22\% = 500 \times 0.22 = 110$$

students in their 4th year. Hence, $\dfrac{30 + 110}{200 + 500} = 0.2 = 20\%$ of the students are in their 4th year.

Answer: 20

Problem 5 Solution

Recall the interior angles of a pentagon add up to $180 \times (5-2) = 540$ degrees, which give us $540 \div 360 = 1.5$ turns around a circle. This means the area of the overlapping circular sectors that are on the vertices of the pentagon add up to the area of 1.5 circles.

Together with the half circle on the side of the pentagon, the areas of all circular sectors that overlap with the pentagon add up to the area of 2 full circles. Thus, the shaded area is $56 - 2 \times 2 = 52$.

Answer: 52

Problem 6 Solution

If we pretend all jellybeans are different, they have $6 \times 9 = 54$ different ways of choosing a pair of jellybeans. There are 2 possible pairs of green jellybeans, 3 possible pairs of red jellybeans, and all other possible pairs are of different colors, thus there are $54 - 2 - 3 = 49$ different pairs of jellybeans of different colors. Therefore, the probability that they pick jellybeans of different colors is $\frac{49}{54}$. So, $P + Q = 49 + 54 = 103$.

Answer: 103

Problem 7 Solution

In 1 minute, Dane can complete $\frac{1}{180}$ of the whole trip and Crystal can complete $\frac{1}{120}$ of the whole trip. This means in one minute they cover $\frac{1}{180} + \frac{1}{120} = \frac{1}{72}$ of the whole trip towards each other. Therefore, if they continue to travel at the same constant speed, in 72 minutes they would meet somewhere on the road.

Answer: 72

Problem 8 Solution

From the original stack of cards every 10th card ends in 0, thus (as $218 \div 10 = 201$ with remainder 8) 201 cards were removed which leaves $2018 - 201 = 1817$ cards remaining.

In this new stack of cards there are 181 cards that end in 0 (as $181 \div 10 = 18$ with remainder 1). Hence after removing them there are $1817 - 181 = 1636$ cards left.

Answer: 1636

Problem 9 Solution

The diagram below shows all possible streets that Quentin could visit while traveling from his house to Renee's house.

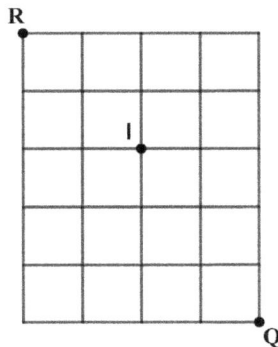

To get to the ice cream shop he must travel 5 blocks: 3 blocks North and 2 blocks West. From the ice cream shop to Renee's house he must travel 4 blocks: 2 blocks North and 2 blocks West.

Any possible path from his house to the ice cream shop can be uniquely determined by a sequence of 5 letters (3 of them Ns and 2 of them Ws), so there are $\binom{5}{2} = 10$ different paths of length 5 from his house to the ice cream shop. Similarly, there are

$\binom{4}{2} = 6$ paths of length 4 from the ice cream shop to Renee's house.

Therefore, there are $10 \times 6 = 60$ different paths of shortest length that go from his house to Renee's stopping at the ice cream shop.

Answer: 60

Problem 10 Solution

Consider the labeled diagram below. We will denote by $[XYZ]$ the area of $\triangle XYZ$.

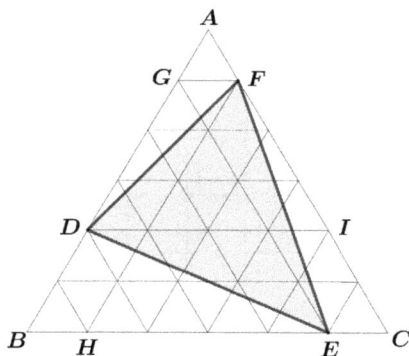

$\triangle FHC$ and $\triangle FEC$ have the same height and their bases are in ratio $5 : 1$, so

$$[FHC] : [FEC] = 5 : 1 = 25 : 5,$$

and $[FEC] = 5$. Similarly,

$$[DIA] : [DFA] = 4 : 1 = 16 : 4$$

and

$$[EGB] : [EDB] = 5 : 2 = 25 : 10,$$

so $[DFA] = 4$ and $[EDB] = 10$.

As $[ABC] = 6 \times 6 = 36$, the shaded region has area $[FDE] = 36 - 4 - 5 - 10 = 17$.

Answer: 17

Problem 11 Solution

We can factor $1440 = 2^4 \times 3^3 \times 5$. Therefore the factors that are perfect cubes are

$$2^0 \times 3^0 \times 5^0 = 1,$$
$$2^3 \times 3^0 \times 5^0 = 8,$$
$$2^0 \times 3^3 \times 5^0 = 27,$$
$$\text{and } 2^3 \times 3^3 \times 5^0 = 216.$$

Thus, there are a total of 4 factors of 2160 that are perfect cubes.

Answer: 4

Problem 12 Solution

Note Wallace must have bought an odd number of boxes of dark chocolate, since the total is an odd number of dollars and the other two kinds of boxes cost an even number of dollars. Furthermore, the cost of a box of white chocolate is the same as the cost of 2 boxes of milk chocolate.

Wallace could have bought 7 boxes of dark chocolate and 1 box of milk chocolate. If he swaps two boxes of dark chocolate for some of the other flavors, he must swap for 1 more box of milk chocolate and 1 of white chocolate. That is, if he bought 5 boxes of dark chocolate, he bought 2 of milk chocolate and 1 of white chocolate.

Following this pattern, we see that if he bought just 1 box of dark chocolate, he must have bought 4 boxes of milk chocolate and 3 boxes of white chocolate.

Answer: 3

Problem 13 Solution

If 1.5 cats eat 1.5 mice in 1.5 hours, then 15 cats eat 15 mice in 1.5 hours. Hence 15 cats can eat $10 \times 15 = 150$ mice in $1.5 \times 10 = 15$ hours.

Answer: 150

Problem 14 Solution

Each trailing zero will come from a factor of $10 = 2 \times 5$. Notice that $320 = 2^6 \times 5$, so if we multiply 320 by a number that has 5^5 as factor, that would give us 5 extra trailing zeros. However, $5^5 = 3125$ is not a 3-digit number.

The biggest power of 5 that is a 3-digit number is $5^4 = 625$, and $625 \times 320 = 200000$. Note any multiple of 625 has more than 3 digits, so this is the only number that gives us 5 trailing zeros when multiplying by 320.

Answer: 625

Problem 15 Solution

To get the maximum number of regions possible, Flo needs to intersect the new circle with all other existing circles. A new circle added to the diagram intersects every existing circle in 2 points, and a new region will be added for each intersection point.

When we have 1 circle there is only 1 region. A second circle intersects the first circle in 2 points, so we have 2 new regions, for a total of $1 + 2 = 3$. A third circle intersects the first two in $2 \times 2 = 4$ points, so the total number of regions increases to $3 + 4 = 7$. Thus, when we add the fourth circle, $3 \times 2 = 6$ new regions will come up, for a total of $7 + 6 = 13$ regions, as shown in the diagram below.

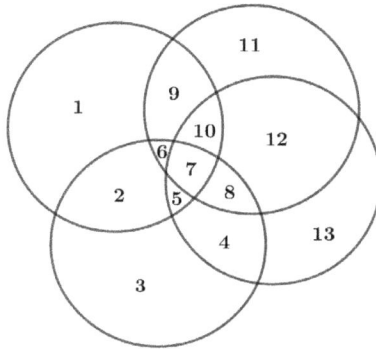

Answer: 13

Problem 16 Solution

To have the smallest possible number of green sport socks in the bag, we will need to have the largest possible number of green dress socks in the bag.

As there is at least one purple dress sock, there can be at most $18 - 1 = 17$ green dress socks in the bag. In this case, there would be $19 - 17 = 2$ pairs of green sport socks in the bag.

Answer: 2

Problem 17 Solution

To reach the other two vertices of the big triangle, Dary will have to walk 20 small paths of length 10 feet, and 20 longer paths of length 10, 20, 30, ..., 190, and 200 feet.Thus, Dary walked

$$(1 + 2 + 3 + \cdots + 19 + 20) \times 10 + 20 \times 10 = 2300$$

feet.

Answer: 2300

Problem 18 Solution

A lizard will flip over every time the number of the minute they are dancing to divides the number of their position on the line. So, they will flip over as many times as factors their position number has.

Note the lizards will stay with their bellies up if they flip over an odd number of times. The only numbers that have an odd number of factors are perfect squares, so only the lizards in positions 1, 4, and 9, so 3 lizards, will stay with their bellies up.

Answer: 3

Problem 19 Solution

Right now the tank is $100\% - 40\% = 60\%$ full. Thus $60\% - 40\% = 20\%$ of the volume of the tank is the same as 40 gallons. Therefore, a full tank is $40 \div 20\% = 40 \div 0.2 = 200$ gallons of water.

Answer: 200

Problem 20 Solution

If we place the black checker piece first, the number of possible places for the white checker piece will depend on what kind of square the black checker piece is.

On the board there are (I) 8 squares that have 4 adjacent squares; (II) 16 squares that have 5 adjacent squares; (III) 4 squares that have 7 adjacent squares; and (IV) 32 squares that have 8 adjacent squares. Examples of each are shown in the diagram below.

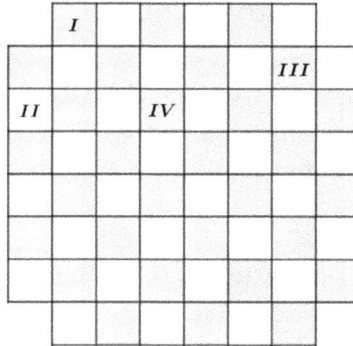

Once we identify how many adjacent squares are to each type of square, we can see there are

$$8 \times 4 + 16 \times 5 + 4 \times 7 + 32 \times 8 = 396$$

different ways of placing the 2 checker pieces so they are adjacent to each other.

Answer: 396

2.7 ZIML April 2018 Division M

Below are the solutions from the Division M ZIML Competition held in April 2018.

The problems from the contest are available on p.65.

Problem 1 Solution
Note we can add one extra small circle in the center:

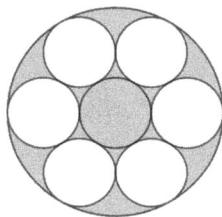

Since all small circles have radius 2, the big circle has radius $3 \times 2 = 6$. Thus, the shaded area is

$$\pi \times 6^2 - 6 \times \pi \times 2^2 = 36\pi - 24\pi = 12\pi.$$

As $\pi \approx 3.14$ we have the area is $12\pi \approx 37.68 \approx 38$ rounded to the nearest integer.

Answer: 38

Problem 2 Solution
From what happened the first day, we see Dixie had a number of candies that leaves a remainder of 3 when divided by $5 + 1 = 6$. The first few numbers that leave a remainder 3 when divided by 6 are

$$9, 15, 21, 27, 33, \ldots.$$

We need to find which of those numbers leaves a remainder of 4 when we divide by 5 after subtracting 3. We see

$$(27 - 3) \div 5 = 4 \text{ with remainder } 4.$$

Therefore 27 is the least number of candies Dixie could have had at the beginning of each day.

Answer: 27

Problem 3 Solution

Since the cookies inside the jars are in ratio $8 : 3$, $\dfrac{8}{11}$ of the cookies went into the Cookie Monster. That means in total she baked $96 \div \dfrac{8}{11} = 96 \times \dfrac{11}{8} = 132$ cookies.

Answer: 132

Problem 4 Solution

As the order in which he chooses the 5 numbers does not matter, there are $\dbinom{15}{5} = 3003$ ways to choose them and, since there are 5 symbols available, there are $3003 \times 5 = 15015$ different ways he can choose the 5 numbers and 1 symbol.

Answer: 15015

Problem 5 Solution

If Mrs. Gatsby only had yellow canaries in her house, there would be a total of $2 \times 116 = 232$ legs and 0 cats legs, so 232 more yellow canary legs than cat legs. That is $232 - 58 = 174$ more canary legs than she should have.

Swapping a yellow canary for a cat changes the difference in the number of legs by $4 + 2 = 6$, thus Mrs. Gatsby has $174 \div 6 = 29$ cats and $116 - 29 = 87$ yellow canaries.

Answer: 87

Problem 6 Solution

The total number of students that came to the park expedition should be divisible by 2, 3, and 5, otherwise the statements about how many students forgot to bring something would not make sense. This means the number of students is a multiple of $2 \times 3 \times 5 = 30$.

We know the number of students is more than 35 and less than 70, so there must have been $30 \times 2 = 60$ students in the expedition to the park.

Answer: 60

Problem 7 Solution

Since we do not want two adults next to each other, we can start by arranging the 6 kids. There are $6! = 720$ ways to do so.

This creates 7 spots that the adults can use to line up for the picture without standing next to each other, so they all just need to choose a different spot. This can be done in $7 \times 6 \times 5 = 210$ different ways.

Thus, there are $720 \times 210 = 151200$ different ways the could line up for the picture.

Answer: 151200

Problem 8 Solution

Each interior angle of an equilateral triangle is $60°$ and each interior angle of a square is $90°$, so the exterior angle formed where two squares and a triangle meet is

$$360° - 90° - 90° - 60° = 120°,$$

which is the same as the interior angle of an hexagon. Thus, upon completion of the figure, a regular hexagon will appear in the middle, with exactly one square on each of its sides and one triangle in each of its vertices.

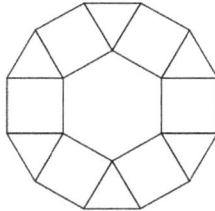

Therefore, the whole figure needed 6 squares and 6 triangles, 12 in total.

Answer: 12

Problem 9 Solution

At a speed of 16 km per hour, it takes Chris $10 \div 16 = 0.625$ hours, or $0.625 \times 60 = 37.5$ minutes to run 10 km.

Since he rests after every 8 minutes of running, he must run/rest 4 full cycles of $8 + 2 = 10$ minutes and then run for an extra $37.5 - 32 = 5.5$ minutes to complete the 10 km. This is $40 + 5.5 = 45.5$ minutes in total.

Answer: 45.5

Problem 10 Solution

The only factors of 2^{2018} that are smaller than 2018 are

$$2^0 = 1, 2^1 = 2, 2^2 = 4, \ldots, 2^9 = 512, 2^{10} = 1024$$

that is, 11 factors ($2^{11} = 2048$ is already too large).

2^{2018} has 2019 factors, so there are $2019 - 11 = 2008$ factors that are larger than 2018.

Answer: 2008

Problem 11 Solution

Notice first that the number of triangles added to the figure increases by 1 each time, and after every group of triangles there is a trapezoid. Thus, the number of figures is such that

$$1 + 1 + 2 + 1 + 3 + 1 + 4 + 1 + \cdots + 1$$
$$= 2 + 3 + 4 + \cdots + 1$$
$$= 35.$$

By following this pattern we can see there are 7 trapezoids and $1 + 2 + \cdots + 7 = 28$ triangles in the figure.

Each trapezoid can be broken down into 3 equilateral triangles, so we can pretend the whole figure was made using $7 \times 3 + 28 = 49$ equilateral triangles. Each equilateral triangle increases the perimeter of the figure by 1, except for the two triangles on the edges, which increase the perimeter by 2.

Thus, the perimeter of the figure is $49 + 2 = 51$.

Answer: 51

Problem 12 Solution

Cincinnati is a word with 10 letters: 2 C's, 3 I's, 3 N's, 1 A, and 1 T. As all the 10 signs looked the same to the crew, they could have arranged them in 10! different ways.

However, as the signs for the repeated letters are indeed identical (even without the plastic), there are $\dfrac{10!}{2! \cdot 3! \cdot 3!} = 50400$ different arrangements for the letters, and only one of them spells Cincinnati correctly, so the probability that they arranged the letters in the correct order is $\dfrac{1}{50400}$. Thus $Q - P = 50400 - 1 = 50399$.

Answer: 50399

Problem 13 Solution

The probability that she goes to the Bahamas is $\dfrac{1}{2}$ and the probability that she goes to Hawaii is also $\dfrac{1}{2}$.

Going to Hawaii she could stay 2 nights if she gets 0 tails, 4 nights if she gets 1 tail, and 6 nights if she gets 2 tails. The probability of getting 1 or 2 tails is $1 - \dfrac{1}{2} \times \dfrac{1}{2} = \dfrac{3}{4}$.

Thus, the probability that she goes to Hawaii and stays 4 or more nights is $\dfrac{1}{2} \times \dfrac{3}{4} = \dfrac{3}{8}$. Therefore, $Q - P = 8 - 3 = 5$.

Answer: 5

Problem 14 Solution

The area where Lady can freely walk can be divided in three regions, each of them a circular sector, as shown on the diagram below.

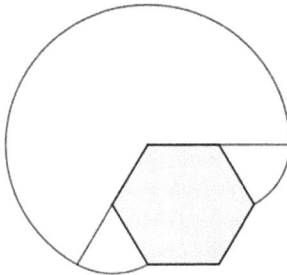

The interior angles of a regular hexagon are $120°$. Therefore the bigger circular sector has radius 4 and an angle of $240°$, so it has area

$$\frac{240}{360} \times \pi \times 4^2 = \frac{2}{3} \times \pi \times 16 = \frac{32\pi}{3}.$$

The two smaller circular sectors are equal, each with radius 2 and angles of $60°$, so they each have area

$$\frac{60}{360} \times \pi \times 2^2 = \frac{1}{6} \times \pi \times 4 = \frac{2\pi}{3}.$$

Altogether the region where Lady can walk has area

$$\frac{32\pi}{3} + 2 \times \frac{2\pi}{3} = 12\pi.$$

Thus, $K = 12$.

Answer: 12

Problem 15 Solution

Recall that the product of two numbers is the same as the product of their GCD and LCM.

We know the product of the numbers is $8400 = 2^4 \times 3 \times 5^2 \times 7$ and their GCD is $20 = 2^2 \times 5$. Thus, both numbers have $2^2 \times 5$ as a factor. This means the numbers could be either $2^2 \times 5 = 20$ and $2^2 \times 3 \times 5 \times 7 = 420$, or $2^2 \times 3 \times 5 = 60$ and $2^2 \times 5 \times 7 = 160$.

As both numbers are greater than 50, they must be 60 and 160. Thus, the smaller number is 60.

Answer: 60

Problem 16 Solution

If Andie had paid full price for all 5 items, he would have paid $75 \times 3 + 25 \times 2 = 275$ dollars.

With the promotions he will only pay $75 \times 2 = 150$ dollars for the jeans and

$$2 \times 25 \times 80\% = 50 \times 0.8 = 40$$

dollars for the shirts, that is, $150 + 40 = 190$ dollars in total.

Thus, Andie saved $275 - 190 = 85$ dollars, so he saved

$$\frac{85}{275} = \frac{17}{55} \approx 30.9\%$$

with his purchase. Therefore $P = 31$.

Answer: 31

Problem 17 Solution
Notice $a + 3b + 3c + d$ is a multiple of 3 if and only if $a + d$ is a multiple of 3, so b and c can be any digit. Thus, we need to count how many pairs of digits (the first of them not 0) add up to a multiple of 3.

If a is one of $1, 2, 4, 5, 7, 8$, there are 3 possible values of d that make $a + d$ a multiple of 3. For example, if $a = 1$, d could be 2, 5, or 8 or if $a = 2$, d could be 1, 4, or 7.

If $a = 3, 6, 9$, there are 4 possible values of d that make $a + d$ a multiple of 3. Thus, there are $6 \times 3 + 3 \times 4 = 30$ ways of choosing a and d, and $10 \times 10 = 100$ ways of choosing b and c.

Therefore, there are $30 \times 100 = 3000$ different 4-digit numbers.

Answer: 3000

Problem 18 Solution
All 4 triangles that make up the shaded region are congruent right triangles which are similar to a triangle with base $2 \cdot 5 = 10$ and height $3 \cdot 5 = 15$.

These similar triangles have a ratio of side lengths of $1 : 6$. Hence the shaded triangles have legs of length $\frac{5}{3}$ and $\frac{5}{2}$, and so each of them has area $\frac{1}{2} \cdot \frac{5}{3} \cdot \frac{5}{2} = \frac{25}{12}$.

Therefore, the area of the whole shaded region is $4 \cdot \dfrac{25}{12} = \dfrac{25}{3}$.
Hence $P + Q = 25 + 3 = 28$.

Answer: 28

Problem 19 Solution
If the number is between 4 and 7, its square will be between 16 and 49. The length of the line segment containing the available numbers is $7.5 - 3 = 4.5$, and the length of the line that contains the numbers between 4 and 7 is $7 - 4 = 3$.

Therefore, the probability is $\dfrac{3}{4.5} = \dfrac{2}{3}$. Thus $Q - P = 3 - 2 = 1$.

Answer: 1

Problem 20 Solution
The speed of the train does not change during the whole journey, and neither does its length. To completely pass through the tunnel/bridge, the train must travel the length of the tunnel/bridge and its own length. It takes the train $32 - 25 = 7$ more seconds to go through the tunnel, than through the bridge, thus the train travels $356 - 230 = 126$ meters in 7 seconds. Hence, the speed of the train is $126 \div 7 = 18$ meters per second.

Since it takes the train 25 seconds to go through the tunnel, it travels $25 \times 18 = 450$ meters through it, which means the train itself is $450 - 230 = 220$ meters long.

The other train approaches at a speed of 22 meters per second, so they get $22 + 18 = 40$ meters closer each second. Each train travels $10 \times 40 = 400$ meters to pass each other, which means the second train is $400 - 220 = 180$ meters long.

Answer: 180

2.8 ZIML May 2018 Division M

Below are the solutions from the Division M ZIML Competition held in May 2018.

The problems from the contest are available on p.75.

Problem 1 Solution

The interior angles of a regular pentagon are $108°$, and the interior angles of a regular hexagon are $120°$. Hence

$$\angle CAB + \angle ABI = 108° + 120° = 228°,$$

and

$$\angle IBC = 360° - 228° = 132°.$$

Since $\triangle BIC$ is isosceles, $\angle BIC = (180° - 132°) \div 2 = 24°$.

Answer: 24

Problem 2 Solution

Since there are 12 students that did not do the English homework and 5 students that did not do English nor Spanish, there are $12 - 5 = 7$ students that did not do English but did the Spanish homework.

This means out of the 34 students in the class, $34 - 7 - 5 - 12 = 10$ students did not do the Spanish homework but did the English homework.

Answer: 10

Problem 3 Solution

Note $6 = 5 + 1 = (1 + 1) \times (2 + 1)$, so a number that has exactly 6 factors is of the form p^5 or pq^2, for p, q prime numbers.

The only number of the form p^5 between 20 and 45 is $2^5 = 32$. The numbers of the form pq^2 between 20 and 45 are $5 \cdot 2^2 = 20$, $7 \cdot 2^2 = 28$, $11 \cdot 2^2 = 44$, and $5 \cdot 3^2 = 45$.

Thus there are 5 numbers with exactly 6 factors.

Answer: 5

Problem 4 Solution

Since the ratio of red pens to black pens is $3 : 11$, for every $3 + 11 = 14$ pens, 3 are red and 11 are black. Thus, Mrs. Darcy has $98 \div 14 \times 3 = 21$ red pens.

Answer: 21

Problem 5 Solution

A number is a multiple of 66 if and only if it is a multiple of 11 and 6.

As we are only using the digits 3 and 6, the number will be a multiple of 6 as long as the last digit is 6.

Remember that a number is a multiple of 11 if and only if the alternating sum of its digits is a multiple of 11. We can think of the alternating sum of a 5-digit number as the difference of its first two digits, plus the difference of the next to digits, plus the last digit. Since the only digits we can use are 3 and 6, the difference of the first digits can be $3 - 3 = 0$, $3 - 6 = -3$, $6 - 3 = 3$, or $6 - 6 = 0$.

Since the last digit will be 6, we need the first two differences to add up to -6 so that the resulting alternating sum is 0 (which is a multiple of 11). To achieve this we need to have $-3 - 3 + 6 = (3 - 6) + (3 - 6) + 6$, so Judy's password is 36366.

Answer: 36366

Problem 6 Solution

Note this is the same as placing 12 identical balls in 5 different boxes, which we can count using stars and bars. We use the non-negative version of stars and bars because it is possible for no students to leave at some of the stops.

Thus, there are $\binom{12+5-1}{5-1} = \binom{16}{4} = 1820$ different lists the bus driver can make.

Answer: 1820

Problem 7 Solution

Label the smaller triangle $\triangle FGE$ as shown below:

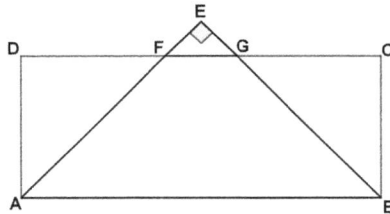

Since $\triangle ABE$ is isosceles, it is a 45-45-90 triangle. Hence in fact $\triangle ADF \cong \triangle BCG$ with both of these triangles being 45-45-90 triangles as well.

Therefore $AD = DF = BC = CG = 2$ and the areas of $\triangle ADF$ and $\triangle BCG$ are both $\frac{1}{2} \cdot 2^2 = 2$. Thus the area of the rectangle outside the triangle is $2 + 2 = 4$.

Answer: 4

Problem 8 Solution

Anastasia can make $3 \times 12 = 36$ tortillas every 20 minutes and Lucresia can make $2 \times 12 = 24$ tortillas every 10 minutes.

During her first 90 minutes Anastasia made $36 \times 90 \div 20 = 162$ tortillas. Hence, when Lucresia joined, they needed to make $456 - 162 = 294$ more tortillas.

Working together they can make $36 + 2 \times 24 = 84$ tortillas in 20 minutes. So they need to work together for $294 \div 84 \times 20 = 70$ minutes.

Answer: 70

Problem 9 Solution

Since A has exactly 5 factors, $A = p^4$ for some prime number p and since B has exactly 6 factors, $B = p^5$ or $B = pq^2$, for some prime numbers p, q.

We can factor $324 = 2^2 \cdot 3^4$. Looking at the exponents of the prime factorization of 324 we see $B = p^5$ is not possible, so $B = pq^2$. Therefore $A = 3^4 = 81$ and $B = 3 \cdot 2^2 = 12$. Thus $A + B = 93$.

Answer: 93

Problem 10 Solution

Whoever works on Friday night will not have to work on Thursday night or Saturday night, so we can start by choosing who will work that night. This can be done in $\binom{5}{2} = 10$ ways.

Then we must choose 2 of the 3 remaining employees to work on Thursday, which can be done in $\binom{3}{2} = 3$ ways.

There is one employee that has not been assigned any night shift yet, so that employee will work on Saturday, and one of the 2 Thursday employees will also work Saturday.

Thus, in total they can choose the weekend shifts in $10 \times 3 \times 2 =$

---

60 different ways.

Answer: 60

Problem 11 Solution

If Alex had gotten only $5 bills, he would have had $32 \times 5 = 160$ dollars in total. That is $360 - 160 = 200$ less dollars than he actually got.

We know that for every $20 bill he got two $10 bills, so we can swap three $5 bills for two $10 bills and one $20 bill and get $2 \times 10 + 20 - 3 \times 5 = 25$ more dollars. To make up the missing $200, we would need to swap bills $200 \div 25 = 8$ times. So he actually had $32 - 3 \times 8 = 8$ $5 bills.

Answer: 8

Problem 12 Solution

The target checkerboard consists of 64 identical squares. Robin's shot determines one row and one column, each containing 8 squares.

Therefore, as one square overlaps, Clinton's shot must land in one of $8 + 8 - 1 = 15$ squares from of the checkerboard. As there are 64 in total, the probability is $\frac{15}{64} \approx 0.234$. As a percent this is $\approx 23.4\%$ so P rounded to the nearest integer is 23.

Answer: 23

Problem 13 Solution

Consider everything in terms of the small triangles.

The full triangle is made up of small triangles, medium triangles made from 4 small triangles, and large triangles made up of $4^2 = 16$ small triangles.

From this we see the full triangle is made of $4 \cdot 16 = 64$ small triangles.

Counting the shaded region is made up of $3 \cdot 1 + 3 \cdot 4 + 1 \cdot 16 = 31$ small triangles. Hence each small triangle has area $93 \div 31 = 3$, so the full triangle has area $64 \cdot 3 = 192$.

Answer: 192

Problem 14 Solution

Note that
$$32 = 5 + 9 + 9 + 9,$$
$$33 = 5 + 5 + 5 + 9 + 9,$$
$$34 = 5 + 5 + 5 + 5 + 5 + 9,$$
$$35 = 5 + 5 + 5 + 5 + 5 + 5 + 5,$$
$$36 = 9 + 9 + 9 + 9.$$

Since any number greater than or equal to 32 is a multiple of 5 more than one of these, it is always possible to pay for any dollar amounts $\geq$ \$32 using only \$5 bills and \$9 bills. However, none of

$$31 - 30 = 1, 31 - 25 = 6, 31 - 20 = 11, 31 - 15 = 16,$$
$$31 - 10 = 21, 31 - 5 = 26, 31 - 0 = 31$$

is a multiple of 9, so it is impossible to pay for \$31 using only these bills without getting any change. Hence 31 is the largest dollar amount impossible to pay for using \$5 and \$9 bills.

Answer: 31

Problem 15 Solution

$\overline{AC}$ is a diagonal, so

$$[ABC] = [ACD] = [ABCD] \div 2 = 120 \div 2 = 60.$$

Since $AE : EF : FC = 1 : 1 : 2$ we have $AE : AC = 1 : 4$ and $FC : AC = 2 : 4 = 1 : 2$.

As the height from B to AC is the height of both $\triangle ABE$ and $\triangle ABC$ we have $[ABE] : [ABC] = AE : AC = 1 : 4$ and hence $[ABE] = [ABC] \div 4 = 60 \div 4 = 15$.

Similarly, $[CFD] : [ACD] = CF : AC = 1 : 2$ and $[CFD] = [ACD] \div 2 = 60 \div 2 = 30$. Therefore $[ABE] + [CFD] = 15 + 30 = 45$.

Answer: 45

Problem 16 Solution

I spent $132 \div 67 = 2$ hours traveling from San Diego to Santa Monica, $200 \div 60 = 3\frac{1}{3}$ hours from Santa Monica to Morro Bay, and $232 \div 58 = 4$ hours from Morro Bay to San Francisco.

Thus I traveled

$$132 + 200 + 232 = 566 \text{ miles}$$

in

$$2 + 3\frac{1}{3} + 4 = 9\frac{1}{3} = \frac{28}{3} \text{ hours.}$$

Therefore my average speed for the whole trip was

$$566 \div \frac{28}{3} = \frac{1698}{28} \approx 60.6 \text{ mph.}$$

Hence rounded to the nearest integer my speed was 61 mph.

Answer: 61

Problem 17 Solution

Note the remainder of 2018^{2018} after dividing by 5 is the same as the remainder of 3^{2018} after dividing by 5. Modulo 5 the first few powers of 3 are

$$3, 4, 2, 1, 3, 4, 2, 1, \ldots$$

which form a cycle of length 4. Since $2018 = 4 \times 504 + 2$, the remainder of 2018^{2018} after dividing by 5 is the second remainder on the list, that is, 4.

Answer: 4

Problem 18 Solution

There are 4 ways to add to 5: $1 + 4$, $4 + 1$, $2 + 3$, or $3 + 2$.

The probability of getting $1 + 4$ is $\left(\dfrac{1}{3}\right)\left(\dfrac{2}{15}\right) = \dfrac{2}{45}$, same as the probability of getting $4 + 1$.

The probability of getting $2 + 3$ is $\left(\dfrac{2}{15}\right)\left(\dfrac{2}{15}\right) = \dfrac{4}{225}$, same as the probability of getting $3 + 2$.

Thus, the probability of the sum being 5 is

$$2 \times \frac{2}{45} + 2 \times \frac{4}{225} = \frac{28}{225}$$

and hence $Q - P = 225 - 28 = 197$.

Answer: 197

Problem 19 Solution

Consider the diagram below:

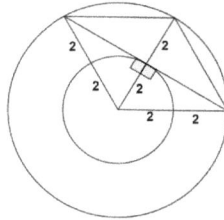

Note the two bottom triangles are right triangles with sides 2 and 4, so they must be 30-60-90 triangles. This similarly implies that the two upper triangles are also 30-60-90 triangles.

Therefore the shaded region is part of a sector with angle $60° + 60° = 120°$. As it is a sector of a circle with radius 4, this sector has an area of

$$\frac{120°}{360°} \cdot \pi \cdot 4^2 = \frac{16\pi}{3}.$$

To find the area of the shaded region, we need to subtract off two of the four triangles. As these two triangles can be rearranged to form one of the equilateral triangles (with side length 4) the total area we need to subtract is

$$\frac{4^2\sqrt{3}}{4} = 4\sqrt{3}.$$

Hence using our approximations the area is

$$\frac{16\pi}{3} - 4\sqrt{3}$$
$$\approx \frac{16 \cdot 3.1}{3} - 4 \cdot 1.7$$
$$= \frac{49.6}{3} - 6.8 \approx 16.53 - 6.8$$
$$= 9.73.$$

Rounded to the nearest tenth is 9.7.

Answer: 9.7

Problem 20 Solution

Buses leave every 35 minutes starting at 5:30 AM, so they leave at 5:30, 6:05, 6:40, 7:15, 7:50, 8:25, etc.

Mr. Lin leaves City A at 7:15, and arrives at City B at 8:10. Therefore he will see all the buses that leave City B by 8:10 and arrive at City A after 7:15.

We see the 6:05 bus from City B arrives at 7:00 in City A, which is before Mr. Lin leaves, but the 6:40 bus from City B will still be traveling. Therefore Mr. Lin sees the buses that leave City B at 6:40, 7:15, and 7:50 (he arrives before the 8:25 bus leaves), a total of 3 buses.

Answer: 3

2.9 ZIML June 2018 Division M

Below are the solutions from the Division M ZIML Competition held in June 2018.

The problems from the contest are available on p.83.

Problem 1 Solution

If Curtis had bought only legal size paper, he would have spent $6.50 \times 28 = 182$ dollars in paper, that is $182 - 150 = 32$ dollars more than he actually spent.

If he swaps 3 packs of legal size paper for 2 regular letter size packs and 1 recycled letter size pack, he would spend

$$3 \times 6.5 - (6 + 6 + 3.5) = 4$$

dollars less. So, he needs to swap packs of paper like this $32 \div 4 = 8$ times. Thus, Curtis bought 8 packs of recycled letter size paper, $2 \times 8 = 16$ packs of regular letter size paper, and $28 - 8 - 16 = 4$ packs of legal size paper.

Answer: 8

Problem 2 Solution

After 4 hours of working she has packed $\dfrac{4}{4+9} = \dfrac{4}{13}$ of the rings.

After packing 135 more, she has packed $\dfrac{9}{13}$ of the rings, that means 135 rings represent $\dfrac{9}{13} - \dfrac{4}{13} = \dfrac{5}{13}$ of the total.

Thus, Esmeralda needs to pack $135 \div \dfrac{5}{13} = 135 \times \dfrac{13}{5} = 351$ rings.

Answer: 351

Problem 3 Solution
Leesa gets 8% of $800, that is

$$800 \times 8\% = 800 \times 0.08 = 64$$

dollars, and 10% of the remaining $1530 - 800 = 730$ dollars, that is,

$$730 \times 10\% = 730 \times 0.1 = 73$$

dollars.

Therefore, Leesa's commission was $64 + 73 = 137$ dollars.

Answer: 137

Problem 4 Solution
Let's pretend Janis traveled 12 miles to get to the top of the hill, so she spent $12 \div 3 = 4$ hours going up, and $12 \div 4 = 3$ hours on her way back.

Therefore, her average speed for the whole trip was

$$24 \div (4 + 3) = 24 \div 7 \approx 3.4$$

miles per hour. Hence rounded to the nearest tenth our answer is 3.4.

Answer: 3.4

Problem 5 Solution
Lilly can make $9 \div 6 = 1.5$ liters of ice cream in one hour, and Jay can make $8 \div 4 = 2$ liters of ice cream in one hour.

Thus, together they can make $1.5 + 2 = 3.5$ liters of ice cream in one hour. This means they need to work together for $21 \div 3.5 = 6$ hours to make 21 liters of ice cream.

Answer: 6

Problem 6 Solution

There are four triangles with base and height 2, two triangles with base and height 1, two triangles with base 1 and height 3, and one square that can be split into 4 small triangles of base and height 1.

Thus the total combined area of all shapes is

$$4 \times \frac{2 \times 2}{2} + 2 \times \frac{1 \times 1}{2} + 2 \times \frac{1 \times 3}{2} + 4 \times \frac{1 \times 1}{2}$$
$$= 8 + 1 + 3 + 2$$
$$= 14$$

Answer: 14

Problem 7 Solution

Note $ABFH$ is an isosceles trapezoid, since $\angle BAH = \angle FBA$ and $AH = BF$. Similarly $DEHG$ is an isosceles trapezoid with

$$\angle AHF = \angle HFB = \angle GHE = \angle DGH$$
$$= \frac{360° - 2 \times (60° + 108°)}{2}$$
$$= 12°.$$

Thus $\angle FHG = 60° - 12° - 12° = 36°$. As $\triangle FGH$ is isosceles,

$$\angle GFH = \angle HGF = \frac{180° - 36°}{2} = 72°.$$

Thus, the smallest angle in $\triangle FGH$ is $36°$.

Answer: 36

Problem 8 Solution

Note the area the paint roller is going to be able to paint each turn is equal to the surface area of the side of the cylinder. The side of the cylinder has area

$$2 \times 3 \times \pi \times 15 \approx 2 \times 3 \times 3.14 \times 15 = 282.6$$

square inches. Since he can turn the roller 5 times before he has to soak it again in pain, he can cover an area of $282.6 \times 5 = 1413$ square inches.

Answer: 1413

Problem 9 Solution

Label the 5 regions A, B, C, D, E as in the diagram below.

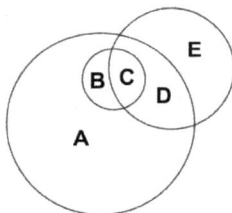

The black area is region A, while the gray area is $C + E$, so we want an expression the same as $A - C - E$.

The circle of radius 3 has area

$$\pi \cdot 3^2 = 9\pi = A + B + C + D.$$

Subtracting the circle of radius 2, with area

$$\pi \cdot 2^2 = 4\pi = C + D + E$$

we get that $9\pi - 4\pi = A + B + E$. This is almost what we want, except we have an extra B and want to subtract a C, which we can do by subtracting the circle of radius 1 with area π.

Therefore
$$A - C - E = 9\pi - 4\pi - \pi = 4\pi$$
is the difference between the black area and the gray area.

Answer: 4

Problem 10 Solution

Note $2\sqrt{3} < 2 \cdot 2 = 4$, so the triangles were drawn on the sides of side length $2\sqrt{3}$.

Thus, the height of each of the triangles is $\dfrac{\sqrt{3}}{2} \times 2\sqrt{3} = 3$. Twice the height of the triangles is equal to the length of the rectangle plus the distance between the vertices. Hence, the distance between the vertices is $2 \times 3 - 4 = 2$.

Answer: 2

Problem 11 Solution

Pretend the students come into the classroom one at a time, and once they come they shake hands with everyone in the classroom before the next student comes in.

The first student to come in does not shake any hands, the second student shakes 1 hand, the third student shakes 2 hands, and the fourth student shakes 3 hands. As this continues, we see that the total number of handshakes is equal to the sum of all positive integers less than the number of students in the class.

Continuing this process we see
$$1 + 2 + \cdots + 13 = 91 \text{ and } 1 + 2 + \cdots + 16 = 136.$$

So, when 136 handshakes occurred, there were 17 students in the classroom and, when 91 handshakes occurred, there were 14 students in the classroom.

Thus, $17 - 14 = 3$ students were missing from class today.

Answer: 3

Problem 12 Solution

$3^3 = 27$ and $3^6 = 729$ and dividing we have $9999 \div 27 = 370$ with remainder 9 and $9999 \div 729 = 13$ with remainder 522.

Thus there are 370 numbers less than 10000 divisible by 3^3 and similarly, 13 numbers less than 10000 divisible by 3^6. We know every number divisible by 3^6 is also divisible by 3^3, hence are $370 - 13 = 357$ numbers divisible by 3^3 but not by 3^6 less than 10000.

Answer: 357

Problem 13 Solution

Using stars and bars we can see there are

$$\binom{10+5-1}{5-1} = \binom{14}{4} = \frac{14!}{4! \times 10!}$$

ways of choosing the flavors of soda. Exactly

$$\binom{10+4-1}{4-1} = \binom{13}{3} = \frac{13!}{3! \times 10!}$$

are such that none of the sodas are Coke and

$$\binom{9+4-1}{4-1} = \binom{12}{3} = \frac{12!}{3! \times 9!}$$

are such that exactly one of the sodas is Coke. Thus, the probability that there are at least two Cokes is

$$1 - \frac{\frac{13!}{3! \times 10!} + \frac{12!}{3! \times 9!}}{\frac{14!}{4! \times 10!}} = 1 - \left(\frac{13!}{3! \times 10!} + \frac{12! \times 10}{3! \times 10!} \right) \times \frac{4! \times 10!}{14!}$$

$$= 1 - \frac{12! \times (13 + 10)}{3! \times 10!} \times \frac{4! \times 10!}{14!}$$

$$= 1 - \frac{23 \times 4}{13 \times 14}$$

$$= 1 - \frac{46}{91}$$

$$= \frac{45}{91}.$$

Therefore $Q - P = 91 - 45 = 46$.

Answer: 46

Problem 14 Solution
The problem only asks about black cards, so we assume there are 5 black cards and 7 other cards.

There are $\binom{5}{3} = 10$ ways to choose the 3 black cards you get

and $\binom{7}{2} = 21$ ways of choosing the remaining two cards. Thus, there are $10 \times 21 = 210$ ways of getting exactly three black cards.

Answer: 210

Problem 15 Solution
After spinning the wheel three times, she can get at ≤ 50 points in one of two ways: she landed on 10 all three times, or if she landed on 10 twice and once on 30 (which can happen in 3 different orders). Any other combination would give her more than 50 points.

Among all circular sectors there are

$$10 + 30 + 40 + 70 + 100 + 50 = 300$$

points, so landing once on 10 has a chance of $\dfrac{10}{300} = \dfrac{1}{30}$ of happening, and landing on 30 has a chance of $\dfrac{30}{300} = \dfrac{1}{10}$ of happening. Thus, the probability that Jessenia wins a major prize is

$$\frac{1}{30} \cdot \frac{1}{30} \cdot \frac{1}{30} + 3 \cdot \frac{1}{10} \cdot \frac{1}{30} \cdot \frac{1}{30} = \frac{10}{27000} = \frac{1}{2700}.$$

Therefore $Q - P = 2700 - 1 = 2699$.

Answer: 2699

Problem 16 Solution
A number that is divisible by 15 must be divisible by 3 and 5. This means the number must end in 5 and the sum of its digits should be a multiple of 3.

If she uses exactly two of each digit, the sum of the digits is $2 + 2 + 3 + 3 + 5 + 5 = 20$, which is not a multiple of 3. Note that using more 3's would make no difference, so she must use more 2's and/or more 5's.

She could either use two more 2's, two more 5's, or one more 2 and one more 5 to have the sum of the digits be a multiple of 3, however, using two 2's will yield the smallest possible number. The smallest number Monica can make using uses four 2's, two 3's, and two 5's is 22223355.

Answer: 22223355

Problem 17 Solution
Note $a + 2b + 3c$ is a multiple pf 3 if and only if $a + 2b$ is a multiple of 3, so c can be any digit.

Thus, we need to count how many pairs of digits (the first of them not 0) are such that $a + 2b$ is a multiple of 3.

If $a = 1, 2, 4, 5, 7, 8$, there are 3 possible values of b that make $a + 2b$ a multiple of 3; if $a = 3, 6, 9$, there are 4 possible values of b that make $a + 2b$ a multiple of 3.

Thus, there are $6 \times 3 + 3 \times 4 = 30$ ways of choosing a and b and, for each of those, 10 ways of choosing c. Therefore there are $30 \times 10 = 300$ such numbers.

Answer: 300

Problem 18 Solution
Since we want only the ones digit, we can ignore the other digits when finding the pattern for the sequence. Starting with 6 the next number is $6 + 3 = 9$. Then the next number is $9 \times 2 = 18$, so we only write down the ones digit: 8.

Continuing we get the sequence

$$6, 9, 8, 1, 2, 5, 0, 3, \ldots$$

Note that after 3 we get back to 6 and the pattern starts over. As 96 is a multiple of 8, the 96th term in the sequence is 3, so we can count to the fourth term in the pattern for the 100th number. Therefore our answer is 1.

Answer: 1

Problem 19 Solution
We know Jerry divides his candy at least between 8 people, otherwise he couldn't have 7 candies left over.

If he had 7 friends, Jerry divided his candy between 8 people today and 7 people yesterday. Therefore the number of candies in his bag must be 4 more than a multiple of 7 and also 7 more than a multiple of 8.

The numbers less than 40 that are 7 more than a multiple of 8 are

$$7, 15, 23, 31, \text{ and } 39,$$

from which 39 is the only one that is 4 more than a multiple of 7. Proceeding similarly for 8, 9 and 10 friends, we see there is no other possible solution. Thus, Jerry share his candies with 7 friends.

Answer: 8

Problem 20 Solution
A door will be locked by as many guards as factors has the number of the cell. So, we need to find how many numbers between 1 and 100 have less than 3 factors.

The only number that has only one factor is 1, and the only numbers that have two factors are the prime numbers. There are 25 prime numbers less than 100, so $1 + 25 = 26$ prisoners will be set free.

Answer: 26

3. Appendix

3.1 Division M Topics Covered

Pre-Algebra and Word Problems*

- Ratios and Proportions: Using ratios to find parts of a whole, Calculating missing information from proportional relationships, Direct and Inverse Proportions, etc.
- Percents: Calculating percent increases and decreases, Relationship between percents and ratios, Using percents in mixture problems (e.g. 40% water and 60% oil)
- Problem Solving Methods: Chicken and Rabbit method, Using ratios when given sums or differences,
- Motion Problems using (Speed)x(Time)=(Distance), Average Speed, Applying direct and inverse proportions to motion problems

- Work using (Rate)x(Time)=(Work Done), Average Rate of Work, Applying direct and inverse proportions to work problems

*Note: Setting up and solving equations is not necessary for any of the problems in Division M. Students are allowed to use equations to solve the questions, but the questions are designed to be solved without using equations or systems of equations.

Geometry

- Areas and Perimeters of Basic Shapes such as triangles, rectangles, parallelograms, trapezoids, and circles
- Angles in Parallel Lines (corresponding angles, alternating interior/exterior angles, same-side interior/exterior angles, etc.)
- Triangles: Congruence and Similarity, Pythagorean theorem, Ratios of Sides for triangles with angles of $45°, 45°, 90°$ or $30°, 60°, 90°$
- Interior and Exterior Angles of Polygons, including the sum of all the interior or exterior angles, the measure of each angle if the polygon is regular, etc.
- Geometric Reasoning with Areas: Congruent shapes have the same area, Similar triangles have a ratio of areas that is the square of the ratio of their sides, Triangles with the same height have a ratio of their areas equal to the ratio of their bases, etc.
- Circles: Arc Length, Sector Area, Definitions for Tangent Lines and Tangent Circles
- Volumes and Surface Areas of Basic Solids such as cubes, spheres, rectangular prisms (boxes), and pyramids

Counting and Probability

- Sum and Product Rules

- Permutations and Combinations
- Counting Methods: Complementary counting, Stars and bars (also called sticks and stones, balls and urns, etc.), Grouping objects that must be together, Inserting objects that must be apart into spaces between objects, etc.
- Sequences: Arithmetic and Geometric Sequences, Sum of elements in an arithmetic sequence, Finding patterns for general sequences
- Probability and Sets: Definitions for event, sample space, complement, intersection, and union, Understanding the use of Venn Diagrams
- Probability in Finite Sample Spaces: Probability as a ratio of outcomes, Probabilities sum to 1, Computing probabilities with complements
- Geometrical Probability: Probability as a ratio of lengths, areas, or volumes
- Basic Statistics: Mean (Average), Median, Mode for lists, Interpreting data from graphs, bar charts, tables, etc.

Number Theory

- Fundamental Definitions: Prime numbers, factors/divisors, multiples, least common multiple (LCM), greatest common factor/divisor (GCF or GCD), perfect squares/cubes/etc.
- Divisibility Rules for numbers such as 2, 3, 4, 5, 8, 9, 10, 11, and how to combine the rules for numbers such as 6, 22, etc.
- (Unique) Prime Factorization and using the prime factorization to find the number of factors, to test whether a number is a perfect square/cube/etc, to find the LCM or GCD, etc.
- Factoring Tricks: Factors come in pairs, perfect squares have an odd number of factors, etc.
- Remainders and Patterns: Finding the units digit, finding the last two digits, finding the remainder when divided by 11, etc.

- Basic Modular Arithmetic: Understand "congruent modulo m" means two numbers have the same remainder when divided by m, The sum of two numbers is congruent modulo m to the sum of the remainders of the numbers when divided by m

3.2 Glossary of Common Math Terms

Acute Angle An angle less than $90°$.

Altitude of a Triangle A line segment connecting a vertex of a triangle to the opposite side forming a right angle. Also called the height of a triangle.

Angle A figure formed by two rays sharing a common vertex. Often measured in degrees.

Arc The curve of a circle connecting two points.

Area The amount of space a region takes up. Often denoted using square brackets: area of $\triangle ABC = [ABC]$.

Arithmetic Sequence A sequence where the difference between one term and the next is constant.

Average See Mean.

Base of a Triangle One side of a triangle, often used when the altitude is drawn from the opposite side to this base.

Binomial Coefficient The symbol $\dbinom{n}{k} = \dfrac{n!}{k!(n-k)!}$.

Chord A line segment connecting two points on the outside of a circle.

Circle A round shape consisting of points that all have the same distance (called the radius) from the center of the circle.

Circumference The perimeter of a circle.

Composite Number A number that is not prime.

Congruent Two shapes or figures that are exactly the same.

Cube A solid figure formed by 6 congruent squares that all meet at right angles.

Deck of Cards A standard deck of cards has 52 cards. There are 4 suits (clubs, diamonds, hearts, and spades) with each suit having cards of 13 ranks (A (ace), $2, 3, \ldots, 10$, J (jack), Q (queen), and K (king)).

Denominator The bottom number in a fraction.

Diagonal A line segment connecting two vertices of a shape or solid that is not an edge of the shape or solid.

Diameter A chord passing through the center of a circle. The diameter has length that is twice the radius.

Die or Dice A standard die (plural is dice) has 6 sides. Each of the 6 sides has the same chance when the die is rolled.

Digit One of $0, 1, 2, \ldots, 9$ used when writing a number.

Distinguishable Objects Objects that are different.

Divisible A number is divisible by another number if there is no remainder when the first number is divided by the second. For example, 35 is divisible by 7.

Divisor A number that evenly divides another number. For example, 6 is a divisor of 48. Also called a factor.

Edge A line segment connecting two vertices on the outside of a shape or solid.

Equally Likely Having the same chance of occurring.

Equiangular Polygon A shape with all equal angles.

Equilateral Polygon A shape with all equal sides.

Equilateral Triangle A regular triangle, one with three equal sides and three equal angles.

Even Number A number divisible by 2.

Exponent The number another number is raised to for powers. For example, in a to the power of b (a^b), the exponent is b.

Face The shape or polygon on the outside of a solid region.

Factor of a Number A number that evenly divides another number. For example, 6 is a factor of 48. Also called a divisor.

Factorial The symbol ! where $n! = n \times (n-1) \times (n-2) \cdots \times 1$.

Fraction An expression of a quotient. For example, $\dfrac{1}{2}$ or $\dfrac{9}{7}$.

Geometric Sequence A sequence where the ratio between one term and the next is constant.

Greatest Common Divisor/Factor (GCD/GCF) The largest number that is a divisor/factor of two or more numbers.

Indistinguishable Objects Objects that are the same.

Intersecting Lines or curves that cross each other.

Intersection of Two Sets The set of objects that are in both of the two sets. Denoted using $\cap$. For example, $\{2,3\} \cap \{3,4,5\} = \{3\}$.

Isosceles Triangle A triangle with two equal sides and two equal angles.

Least Common Multiple (LCM) The smallest number that is a multiple of two or more numbers.

Mean The sum of the numbers in a list divided by the how many numbers occur in the list. Also called the average.

Median The number in the middle of a list when the list is arranged in increasing order.

Midpoint The point in the middle of a line segment.

Mode The number or numbers occurring most often in a list of numbers.

Multiple A number that is an integer times another number. For example, 72 is a multiple of 8.

Numerator The top number in a fraction.

Obtuse Angle An angle between $90°$ and $180°$.

Odd Number A number not divisible by 2.

Parallel Lines Lines that do not intersect.

Perfect Cube A number that is another number cubed. For example, $64 = 4^3$ is a perfect cube.

Perfect Square A number that is another number squared. For example, $64 = 8^2$ is a perfect square.

Perimeter The length/distance around the outside of a shape.

Pi (π) A number used often in geometry. $\pi = 3.1415926\dots \approx$ $3.14 \approx \dfrac{22}{7}$.

Polygon A shape formed by connected line segments.

Prime Factorization The expression of a number as the product of all its prime factors. For example, 24 has prime factorization $2 \times 2 \times 2 \times 3 = 2^3 \times 3$.

Prime Number A number whose only factors are one and itself.

Proportional Ratios Ratios that have equal values when expressed in fraction form. For example, $2 : 3$ is proportional to $8 : 12$.

Quadrilateral A shape with four sides.

Quotient The integer quantity when dividing one number by another. For example, the quotient of $38 \div 5$ is 7 as $38 = 7 \times 5 + 3$.

Radius of a Circle The distance from the center of the circle to any point on the outside of the circle.

Randomly Chosen for a group of objects. Unless specified, the chance of choosing each object is the same as any other object.

Rank of a Card See Deck of Cards.

Ratio A relation depicting the relation between two quantities. For example $2 : 3$ or $\dfrac{2}{3}$ denotes that for every 3 of the second quantity there are 2 of the first quantity.

Rational Number A number that can be written as a fraction.

Reciprocal One divided by the number. For example, the reciprocal of 7 is $\frac{1}{7}$.

Rectangle A quadrilateral with four right angles (an equiangular quadrilateral).

Regular Polygon A polygon with all equal sides and all equal angles (equilateral and equiangular).

Remainder The quantity left over when one integer is divided by another. For example, the remainder of $38 \div 5$ is 3 as $38 = 7 \times 5 + 3$.

Rhombus A quadrilateral with four equal sides (an equilateral quadrilateral).

Right Angle A $90°$ angle.

Right Triangle A triangle containing a right angle.

Scalene Triangle A triangle with three unequal sides and three unequal angles.

Sector The region formed by an arc and the two radii connecting the ends of the arc to the center of the circle.

Sequence An ordered list of numbers.

Set An unordered collection or group of objects without repeated elements. Denoted using curly brackets. For example, $\{1, 2, 3, 4\}$ is the set containing the integers $1, \ldots, 4$.

Similar Shapes or solids that have the same angles and sides that share a common ratio.

Simplest Radical Form An expression containing a radical such that the number inside the radical is an integer that has no perfect squares.

Sphere A round solid consisting of points that all have the same distance (called the radius) from the center of the sphere.

Square A shape with four equal sides and four equal angles (a regular quadrilateral).

Subset A set of objects that is contained inside a larger set of objects. Denoted using $\subseteq$. For example $\{2,3\} \subseteq \{1,2,3,4\}$.

Suit of a Card See Deck of Cards.

Surface Area The total area of all the faces of a solid.

Trapezoid A quadrilateral with one pair of parallel sides.

Triangle A shape with three sides.

Union of Two Sets The set of objects that are in one or both of the two sets. Denoted using $\cup$. For example, $\{2,3\} \cup \{3,4,5\} = \{2,3,4,5\}$.

Venn Diagram A diagram with circles used to understand the relationship between overlapping sets.

Vertex The intersection of line segments, especially the intersection of sides or edges in a shape or solid.

Volume The amount of space a solid region takes up.

With Replacement When choosing objects with replacement, a chosen object is returned to the others allowing it to be chosen more than once.

3.3 ZIML Answers

ZIML October 2017 Division M

Problem 1:	30	**Problem 11:**	13
Problem 2:	144	**Problem 12:**	45
Problem 3:	66	**Problem 13:**	3600
Problem 4:	14.1	**Problem 14:**	82560
Problem 5:	989	**Problem 15:**	7
Problem 6:	45	**Problem 16:**	15
Problem 7:	1	**Problem 17:**	1260
Problem 8:	7	**Problem 18:**	6
Problem 9:	32	**Problem 19:**	18
Problem 10:	42	**Problem 20:**	30240

ZIML November 2017 Division M

Problem 1:	392	**Problem 11:**	15
Problem 2:	12	**Problem 12:**	13
Problem 3:	2	**Problem 13:**	7
Problem 4:	2	**Problem 14:**	64
Problem 5:	70	**Problem 15:**	90720
Problem 6:	12	**Problem 16:**	32857
Problem 7:	180	**Problem 17:**	252
Problem 8:	60	**Problem 18:**	967
Problem 9:	32	**Problem 19:**	75
Problem 10:	540	**Problem 20:**	150

ZIML December 2017 Division M

Problem 1:	5	**Problem 11:**	969
Problem 2:	6	**Problem 12:**	20
Problem 3:	48	**Problem 13:**	78
Problem 4:	760	**Problem 14:**	13
Problem 5:	48	**Problem 15:**	4.8
Problem 6:	16	**Problem 16:**	38
Problem 7:	21	**Problem 17:**	1.7
Problem 8:	218	**Problem 18:**	5000
Problem 9:	12	**Problem 19:**	0
Problem 10:	7	**Problem 20:**	44100

ZIML January 2018 Division M

Problem 1:	76	**Problem 11:**	57
Problem 2:	81	**Problem 12:**	6.25
Problem 3:	6	**Problem 13:**	89
Problem 4:	3.4	**Problem 14:**	901263
Problem 5:	16	**Problem 15:**	8
Problem 6:	57	**Problem 16:**	10
Problem 7:	27	**Problem 17:**	14
Problem 8:	4	**Problem 18:**	53
Problem 9:	6.5	**Problem 19:**	3750
Problem 10:	9870	**Problem 20:**	6

ZIML February 2018 Division M

Problem 1:	81	**Problem 11:**	48
Problem 2:	2	**Problem 12:**	1261
Problem 3:	52341	**Problem 13:**	6
Problem 4:	9	**Problem 14:**	56
Problem 5:	6	**Problem 15:**	15
Problem 6:	981	**Problem 16:**	331776
Problem 7:	2	**Problem 17:**	45
Problem 8:	59	**Problem 18:**	7
Problem 9:	60	**Problem 19:**	8
Problem 10:	24	**Problem 20:**	480

ZIML March 2018 Division M

Problem 1:	2	Problem 11:	4
Problem 2:	8	Problem 12:	3
Problem 3:	8	Problem 13:	150
Problem 4:	20	Problem 14:	625
Problem 5:	52	Problem 15:	13
Problem 6:	103	Problem 16:	2
Problem 7:	72	Problem 17:	2300
Problem 8:	1636	Problem 18:	3
Problem 9:	60	Problem 19:	200
Problem 10:	17	Problem 20:	396

ZIML April 2018 Division M

Problem 1:	38	Problem 11:	51
Problem 2:	27	Problem 12:	50399
Problem 3:	132	Problem 13:	5
Problem 4:	15015	Problem 14:	12
Problem 5:	87	Problem 15:	60
Problem 6:	60	Problem 16:	31
Problem 7:	151200	Problem 17:	3000
Problem 8:	12	Problem 18:	28
Problem 9:	45.5	Problem 19:	1
Problem 10:	2008	Problem 20:	180

ZIML May 2018 Division M

Problem 1:   24

Problem 2:   10

Problem 3:   5

Problem 4:   21

Problem 5:   36366

Problem 6:   1820

Problem 7:   4

Problem 8:   70

Problem 9:   93

Problem 10:   60

Problem 11:   8

Problem 12:   23

Problem 13:   192

Problem 14:   31

Problem 15:   45

Problem 16:   61

Problem 17:   4

Problem 18:   197

Problem 19:   9.7

Problem 20:   3

ZIML June 2018 Division M

Problem 1:	8	**Problem 11:**	3
Problem 2:	351	**Problem 12:**	357
Problem 3:	137	**Problem 13:**	46
Problem 4:	3.4	**Problem 14:**	210
Problem 5:	6	**Problem 15:**	2699
Problem 6:	14	**Problem 16:**	22223355
Problem 7:	36	**Problem 17:**	300
Problem 8:	1413	**Problem 18:**	1
Problem 9:	4	**Problem 19:**	8
Problem 10:	2	**Problem 20:**	26

www.ingramcontent.com/pod-product-compliance
Lightning Source LLC
Chambersburg PA
CBHW060553210326
41519CB00014B/3458